EXERCISES FOR WEATHER & CLIMATE

SEVENTH EDITION

GREG CARBONE
University of South Carolina–Columbia

Prentice Hall

New York Boston San Francisco
London Toronto Sydney Tokyo Singapore Madrid
Mexico City Munich Paris Cape Town Hong Kong Montreal

Editor: *Christian Botting*
Project Manager: *Tim Flem*
Assistant Editor: *John DeSantis*
Editor-In-Chief, Chemistry and Geosciences: *Nicole Folchetti*
Managing Editor, Science: *Gina M. Cheselka*
Project Manager, Science: *Wendy A. Perez*
Cover Designer: *Maureen Eide*
Compositor: *Karen Beidel*
Senior Media Producer: *Angela Berhardt*
Operations Specialist: *Amanda Smith*
Cover Photo Credit: © NOAA - Corbis

Pearson Prentice Hall
Pearson Education, Inc.
Upper Saddle River, New Jersey 07458

ISBN-10: 0-321-59625-0
ISBN-13: 978-0-321-59625-3

Printed in the United States of America
10 9 8 7 6 5 4 3

Prentice Hall
is an imprint of

www.pearsonhighered.com

Contents

Preface

This lab manual presents exercises for introductory weather and climate students. Its questions are designed to encourage critical thinking about atmospheric processes through data analysis, problem solving, and experimentation.

The manual was written to complement two Pearson Prentice Hall weather and climate textbooks—Aguado and Burt's *Understanding Weather and Climate,* and Lutgens and Tarbuck's *The Atmosphere.* However, it should be appropriate for most introductory atmospheric science courses. I hope that the exercises help students apply what they've learned in lectures and their text. A few questions in each lab may require reference to these resources, but most can be answered from material contained within the manual

Instructors who have used previous editions will find minor editorial changes scattered through this edition. Four labs underwent more extensive changes. Lab 3 retains the same approach as in the sixth edition but uses new data sets to illustrate shortwave and long-wave radiation fluxes; it also adds a series of questions that explore sensible and latent heat fluxes. It retains interactive computer modules on shortwave and long-wave radiation. These, and most of the interactive exercises have been completely repackaged as Flash animations to offer improved functionality and a more consistent style across seven interactive modules. Lab 11 adds two recent tornado case studies and associated synoptic weather maps. Labs 4 and 16, focusing on the earth's energy budget and forcing factors of climate change, formerly used an energy balance model developed by James Burt. This interactive model survived more than two decades of Microsoft operating systems changes, but because the code does not run properly in Vista, the model is not packaged with this edition of the manual, and both labs have been reworked. Lab 4 uses output from Burt's one-dimensional energy balance model to illustrate spatial and seasonal variations in solar radiation, albedo, poleward energy transport, and temperature. It concludes with a discussion of other temperature controls. Lab 16 uses output from the energy balance model to simulate the effects of solar variability, aerosols, and orbital changes. The lab also uses output from the Model for the Assessment of Greenhouse-Gas Induced Climate Change (MAGICC) model developed by Tom Wigley, Sarah Raper, Mike Salmon, and Tim Osborn.

Many have contributed to this seventh edition. Professors Mace Bentley, Georgina DeWeese, Jamie Dyer, Jay Martinelli, Richard Schultz, and Thomas Williams all provided constructive suggestions through the review process. Teaching assistants and students at the University of South Carolina continue to provide feedback graciously. Dan Kaveney, Tim Flem, and Wendy Perez at Pearson Prentice Hall guided the revision. Angela Bernhardt at Pearson Prentice Hall shepherded the conversion and redesign of the interactive software that was accomplished by Chris Andreola and his colleagues at adcSTUDIO. Karen Beidel edited text, constructed graphics, and did page layout. My thanks to all of them.

Software Notes

Minimum System Requirements

WINDOWS
- Windows XP, Vista
- 1024 x 768 screen resolution
- Internet Explorer 6, 7
- Firefox 2.0, 3.0

MACINTOSH
- Mac OS 10.3, 10.4, 10.5
- 1024 x 768 screen resolution
- Safari 1.3, 2.0, 3.0
- Firefox 2.0, 3.0

Color monitor running "Thousands of Colors" or higher (16bit or higher)
4x CD-ROM drive

This CD is designed for stand-alone use and is not engineered or supported for use over a network.

Installation Instructions

Windows
- If the application does not start automatically, locate the CD drive on your computer and double-click the “StartHere” weather icon.

Macintosh
- Locate the Carbone7e CD and double-click the “StartHere” weather icon.

Lab 1

VERTICAL STRUCTURE OF THE ATMOSPHERE

Materials Needed

- calculator
- ruler

Introduction

Our first lab introduces the concept of atmospheric pressure. We will construct and interpret a number of graphs to measure how pressure, density, and temperature change with height above the earth's surface. We will focus on how these relationships are expressed in the troposphere, which is where most weather processes occur.

Changes in Atmospheric Pressure with Height

The atmosphere is a compressible fluid, made up of gases whose molecules are pulled to the earth's surface by gravity. As a result, the molecules that make up the atmosphere are most compressed close to the earth's surface, and atmospheric density decreases most rapidly with height there (Figure 1-1).

Although the boundary between the earth's surface and the atmosphere is obvious, there is no clear "top" to the atmosphere. It thins out with increasing height, but never actually ends. (The phenomenon is analogous to repeatedly dividing a number in half. Each division produces a smaller number, but theoretically one never reaches zero.) However, since very few gas molecules within earth's gravitational field exist beyond 100 kilometers (km), we can consider this height an arbitrary "top" to the atmosphere.

We may use a simple rule to describe the rate at which density decreases with height: for every 5.6 km you ascend, there is half the atmospheric mass above you as when you started.

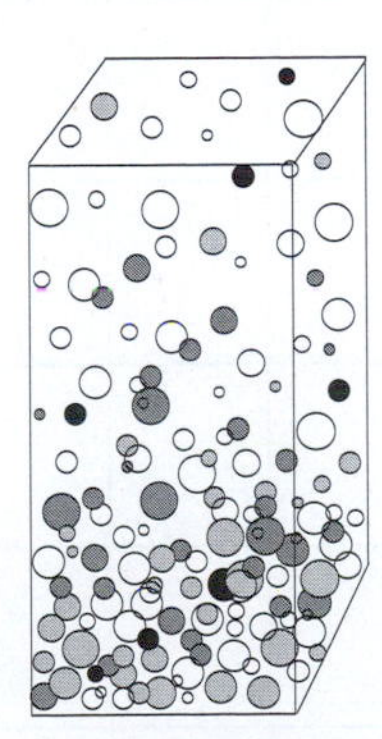

Figure 1-1

1. *Using the above rule, indicate the percentage of the atmosphere above each height in Table 1-1.*

2. *Use the data in Table 1-1 to construct a graph below. The vertical axis is divided into 12 equally spaced intervals. Label this axis "Height above Sea Level (km)" and label the intervals. Label the horizontal axis "Percentage of the Atmosphere Above" and label its intervals (0–100%).*

Table 1-1

Height (km)	% Above
22.4	
16.8	
11.2	
5.6	
Sea level	100%

Since barometric pressure reflects the weight of the atmosphere above a point, there is also a close relationship between height and atmospheric pressure. We can assume that 100% of the atmospheric mass lies above sea level and exerts a pressure of approximately 1000 millibars (mb). Since atmospheric mass at 5.6 km is 50% of its sea-level value, the pressure at this height is half of that exerted at sea level.

3. *Using this relationship, add a pressure scale beneath the percent scale in your graph and label it "Pressure (mb)."*

You can estimate the pressure level for specific heights using a mathematical form of our rule of thumb:

$$\text{Pressure} = 1000 \text{ mb} \bullet (0.5)^{x/5.6}$$

where x = height in kilometers

For example, at the top of Mt. Everest (8.85 km, 29,035 ft):

$$\begin{aligned}\text{Pressure} &= 1000\text{mb} \bullet (0.5)^{8.85/5.6} \\ &= 1000 \text{ mb} \bullet (0.5)^{1.58} \\ &= 1000 \text{ mb} \bullet (0.33) \\ &= 330 \text{ mb}\end{aligned}$$

4. *Using this equation or your graph, estimate the percentage of the atmosphere above and the total pressure for the three heights indicated.*

- *The height of a cruising jet (11.2 km, 37,000 ft)*
  __________ *% above* __________ *mb*

- *The top of Mt. McKinley (6.19 km, 20,320 ft)*
  __________ *% above* __________ *mb*

- *The top of Pike's Peak (4.34 km, 14,110 ft)*
  __________ *% above* __________ *mb*

- *Surface*
  100 % above 1000 mb

Total atmospheric pressure, according to *Dalton's law,* is the sum of the partial pressures that each gas in the atmosphere exerts. For example, since nitrogen makes up 78% of the total atmospheric volume, its partial pressure is approximately 780 mb at sea-level. Oxygen, the other major component of the atmosphere, makes up 21% of the atmosphere's volume.

5. *What is the approximate partial pressure of oxygen at sea level?*

The partial pressure exerted by each gas decreases with height at the same rate as total atmospheric pressure does. (The proportion of each gas remains fixed in the lowest 80 kilometers of the atmosphere.) Since at 5.6 km the total atmospheric pressure drops to 500 mb, i.e., 50% of its sea-level value, the partial pressure of nitrogen at 5.6 km is approximately 390 mb, also 50% of its sea-level value.

6. *What is the partial pressure and percentage of sea-level oxygen at the top of*

- *Pike's Peak* __________ *mb* __________ *%*
- *Mt. Everest* __________ *mb* __________ *%*

Your answers above should help you understand why physical activity is more taxing at higher elevations and why jets have pressurized cabins.

Changes in Temperature with Height

We define four layers of the atmosphere (the *troposphere, stratosphere, mesosphere,* and *thermosphere*) according to their average lapse rate—the rate at which temperature changes with height (Figure 1-2). Although the lapse rate at any given time or place will differ from this average, the figure provides a starting point for understanding the temperature profile of the atmosphere. The tropopause, stratopause, and mesopause mark the top (end) of each layer.

7. *Ozone is a good absorber of ultraviolet radiation from the sun. How does its highest concentration at 20–30 km influence the stratospheric lapse rate?*

Tropospheric Lapse Rate

Let us examine in more detail the environmental lapse rate in the troposphere—the layer in which most weather phenomena occur. When averaged over all seasons, air temperature is 15°C at the earth's surface and decreases by 6.5°C per kilometer in the lowest 11 km. Consequently, this decrease is often referred to as the *average lapse rate*—the average for all locations and seasons. The resulting temperatures characterize the "standard atmosphere" shown in Table 1-2.

8. *Fill in the temperature values in Table 1-2 for the standard atmosphere from 2,000 to 10,000 meters.*

Of course, the measured atmospheric lapse rate for a specific time and place will likely differ from the average.

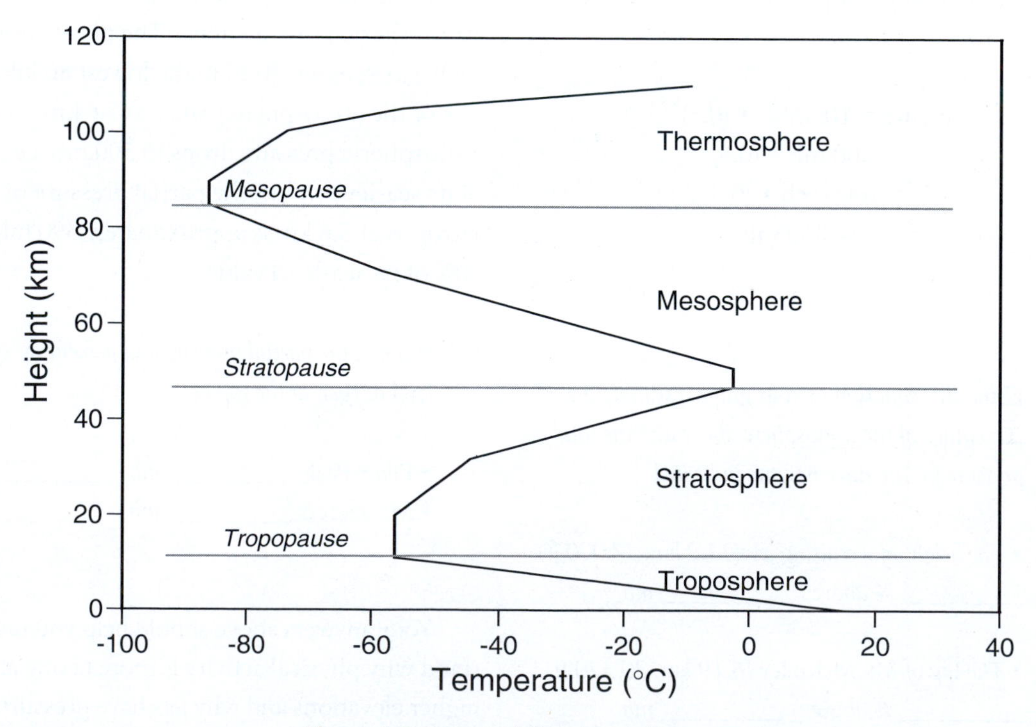

Figure 1-2. Layers of the atmosphere.

Table 1-2

Standard Atmosphere

Height (m)	sea level	1000	2000	4000	6000	8000	10000	11000	14000	16000
Temperature (°C)	15.0	8.5	______	______	______	______	______	-56.5	-56.5	-56.5
Pressure (mb)	1013.25		795			357				
Density (kg m^{-3})	1.23		1.01			0.35				

Key West, Florida

Height (m)	surface	2000	4000	6000	8000	10000	12000	14000	16000	18000
Temperature (°C)	27.2	15.2	6.1	-6.7	-17.9	-32.6	-50.5	-64.5	-75.6	-73.3
Pressure (mb)		800	630		378	286				
Density (kg m^{-3})		0.97			0.52					

Fairbanks, Alaska

Height (m)	surface	400	1000	2000	4000	6000	8000	10000	12000	14000
Temperature (°C)	0.4	5.4	0.9	-4.4	-17.5	-32.8	-45.7	-52.7	-49.7	-52.9
Pressure (mb)			893	782	600		333			
Density (kg m^{-3})				1.02			0.51			

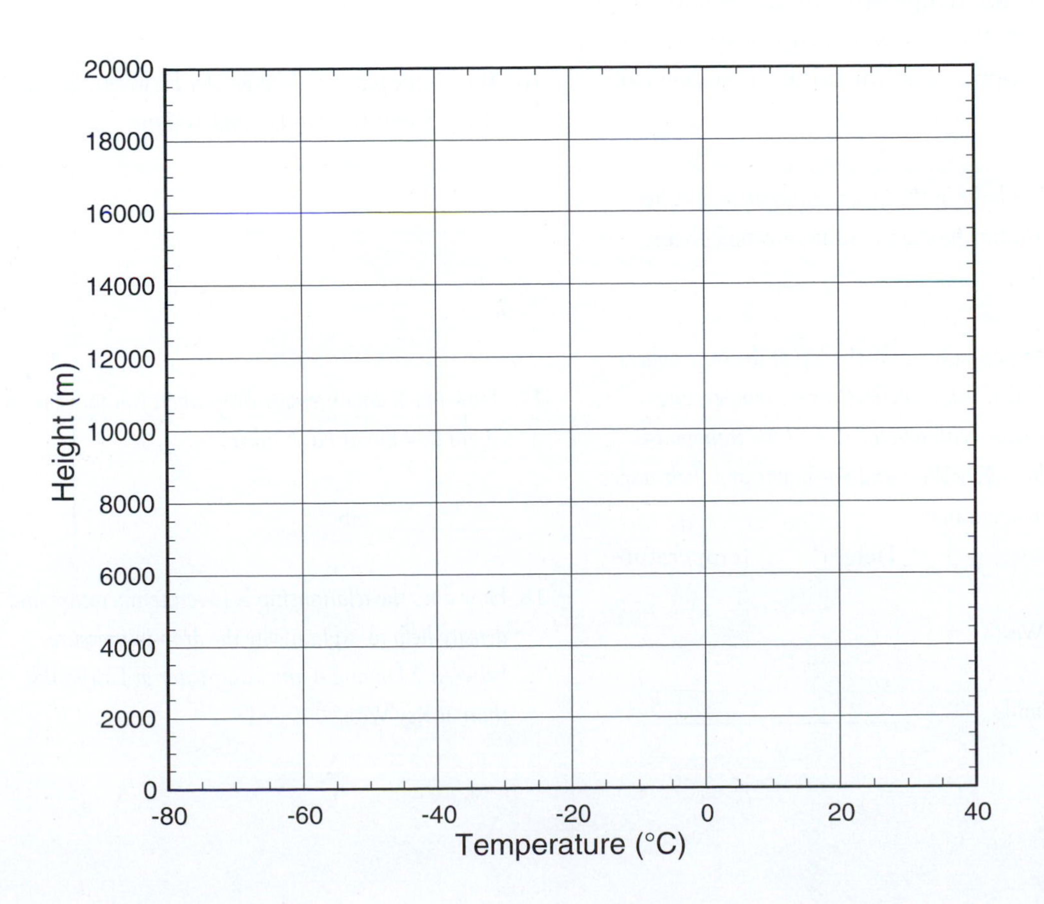

9. *Table 1-2 also lists atmospheric temperatures measured at various heights above Key West, Florida, and Fairbanks, Alaska, on October 5, 2005, 1200 Greenwich Mean Time (GMT). Plot the temperature values for these two stations and the standard atmosphere on the graph beneath the table.*

10. *Which station has the higher surface temperature?*

    Which station has the higher temperature at 10 km?

    At 14 km?

As your graph shows, there are some situations in which temperature increases with height in certain layers of the troposphere. Since this is the opposite of the norm, such layers are called *inversion layers.*

11. *Circle a layer in the lower troposphere of either temperature profile where an inversion occurs.*

12. *The tropopause marks the top of the troposphere and is defined as the level where temperature ceases to decrease with height. Record the tropopause height at Key West and Fairbanks and their respective temperatures.*

	Height	Temperature
Key West	_________	_________
Fairbanks	_________	_________

13. *What relationship do you see between average tropospheric temperature and the height of the tropopause? If vertical mixing of air is responsible for a thicker troposphere, what does our example say about the relationship between temperature and vertical mixing in the troposphere?*

14. *How much does pressure drop when you move from 2 km to 4 km at Key West?*

    _________ *mb*

15. *How much does pressure drop when you move from 8 km to 10 km at Key West?*

    _________ *mb*

16. *Why is the pressure drop greater between 2 km and 4 km than between 8 km and 10 km?*

17. *How much does pressure drop when you move from 2 km to 4 km at Fairbanks?*

    _________ *mb*

18. *How does the relationship between temperature and density help to explain why the drop in pressure between 2 km and 4 km was greater at Fairbanks than at Key West?*

Review Questions

In your own words, describe how air pressure, density, and temperature vary with height in the troposphere.

The thickness of the troposphere varies from place to place and from day to day. What influences this thickness?

We often hear the expression, "cold air sinks." If so, why isn't the colder air of the upper and middle troposphere steadily flowing down to the surface?

In which case do you think you would measure a greater pressure change: moving a given distance horizontally or moving the same distance vertically? (As a reference, an average hurricane has a radius of approximately 600 km and a central pressure of 950 mb.)

Lab 2

Earth–Sun Geometry

Materials Needed

- calculator (with trigonometric functions)
- ruler

Introduction

What causes the seasons and changes in the amount of daylight we receive? The diagrams and exercises in this lab show how earth–sun geometry influences these variables. We examine the earth–sun relationship early in our study of weather and climate because most atmospheric processes are ultimately driven by spatial variations in solar energy.

Earth–Sun Relationships

The distance between the earth and the sun averages about 150 million kilometers (93 million miles). Because of this distance and the earth's relatively small size compared with that of the sun, it is reasonable to assume that the sun's rays strike the nearly spherical earth in straight paths.

The earth's axis of rotation is tilted 23½° from the perpendicular to the plane of the ecliptic—the plane on which the earth revolves around the sun. This tilt is oriented in the same direction throughout the year, with the North

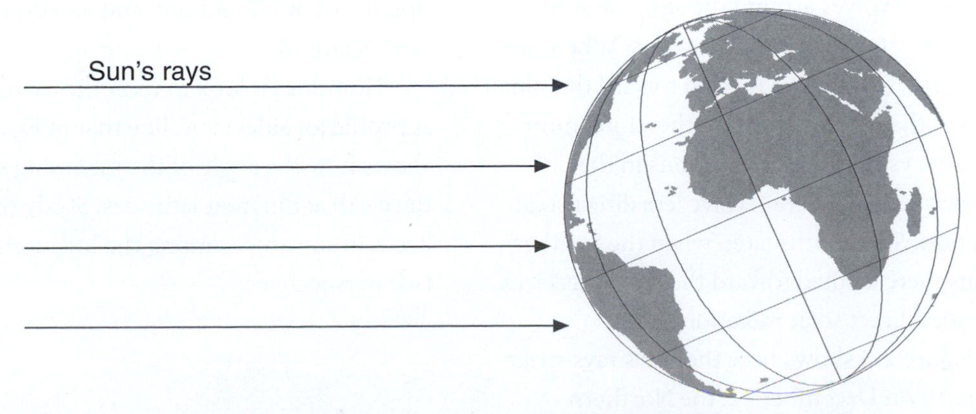

Figure 2-1. Parallel rays striking the spherical earth.

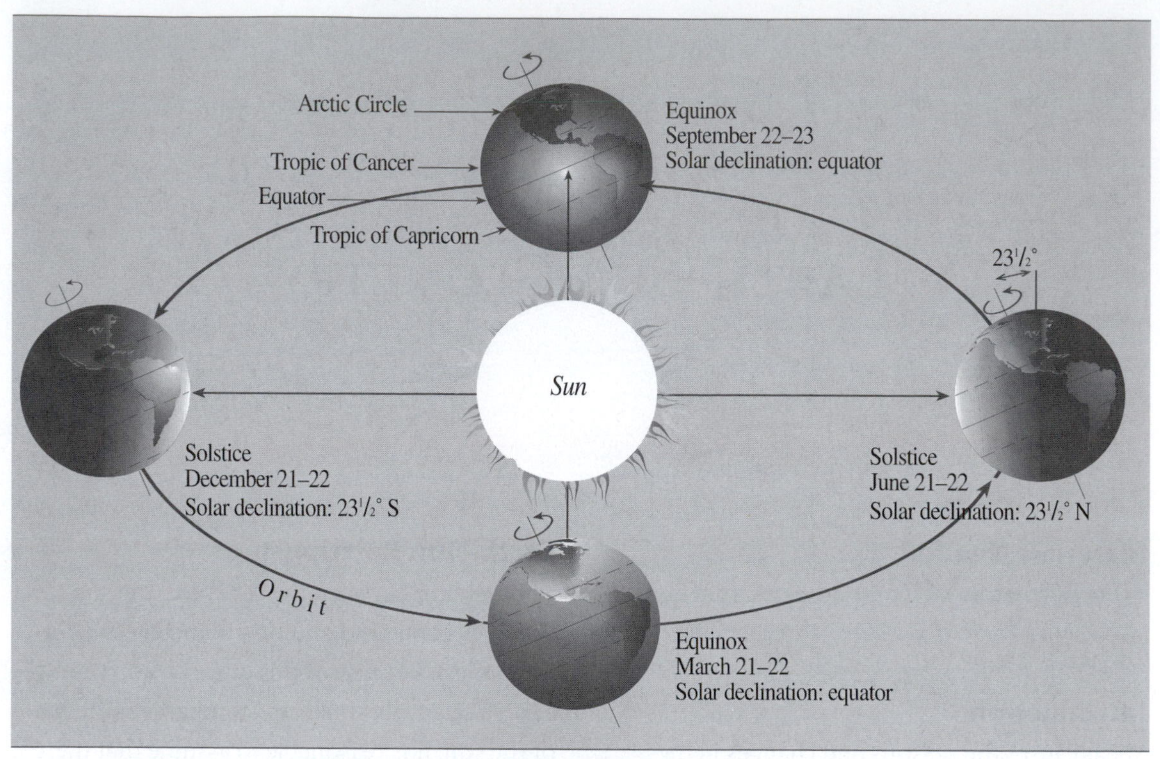

Figure 2-2. The earth's revolution around the sun. Notice that the axis tilts in the same direction throughout the year.

Pole presently pointing toward the North Star, Polaris. Figure 2-2 (not to scale) shows that the Northern Hemisphere is tilted toward the sun during its summer months and away from the sun during its winter months.

Our seasons occur because of this tilt. As the earth revolves around the sun, the sun's direct rays strike different latitudes. When the Northern Hemisphere is tilted toward the sun, it receives the more direct and, therefore, more intense rays of the sun. Locations in the Southern Hemisphere receive less direct solar radiation. Six months later, when the Southern Hemisphere is tilted toward the sun, it receives the more direct solar radiation.

Figure 2-3 shows how the sun's rays strike the earth on December 22, the Northern Hemisphere's winter solstice. At solar noon on this date, the sun's rays are perpendicular to the earth's surface at 23½° S (location D). As we move away from 23½ ° S we see that the rays of the sun strike the earth's surface at progressively lower angles. Location C is at the equator, location B is at 30° N latitude, and location A is at 66½° N latitude.

How does it look to us at the earth's surface? A profile (or side) view, like that in Figure 2-4, shows how the angle of the incoming sun strikes the earth at different latitudes. Study the differences in sun angle among the locations from the two perspectives.

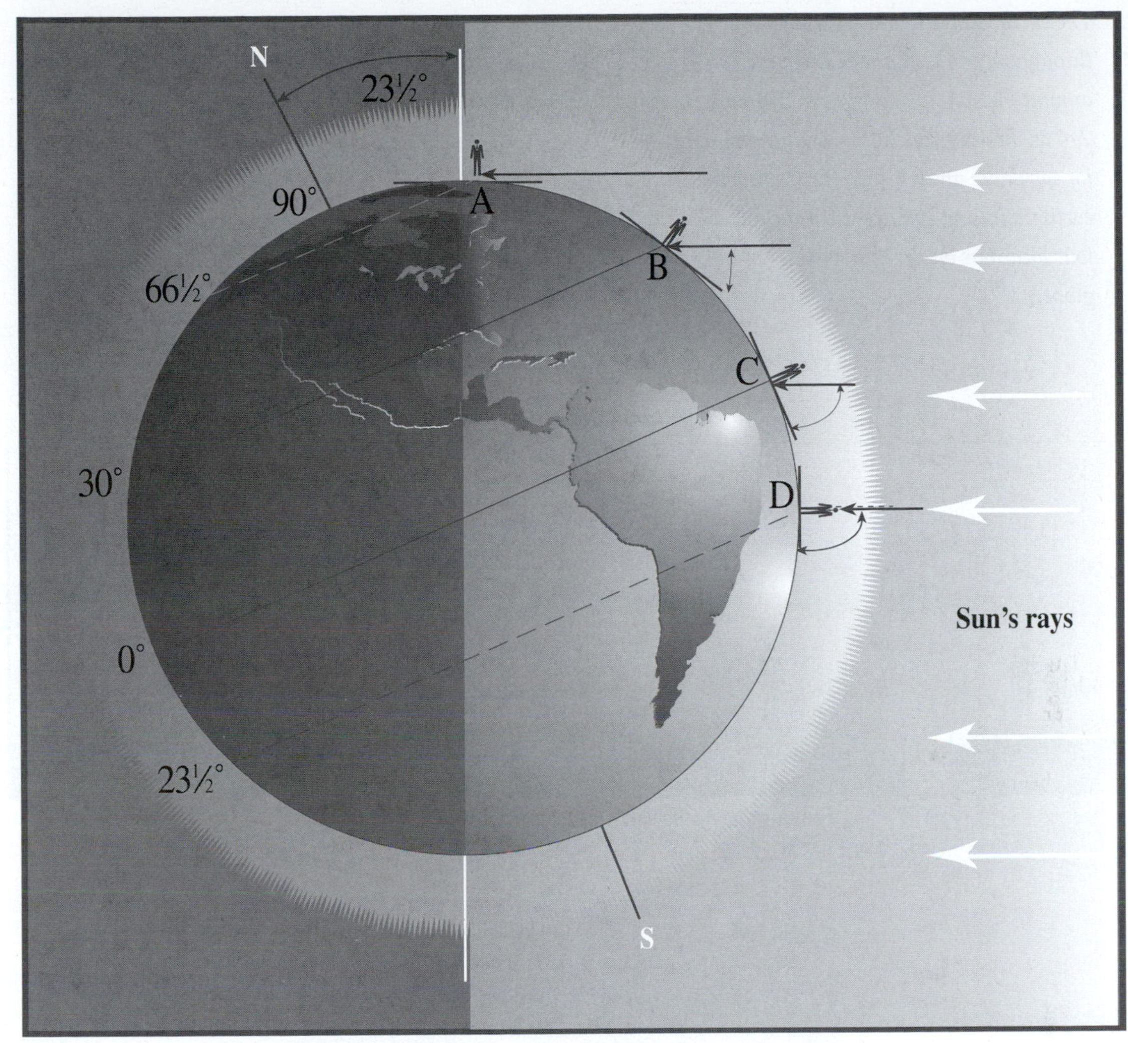

Figure 2-3. Sun's rays striking the earth on December 22.

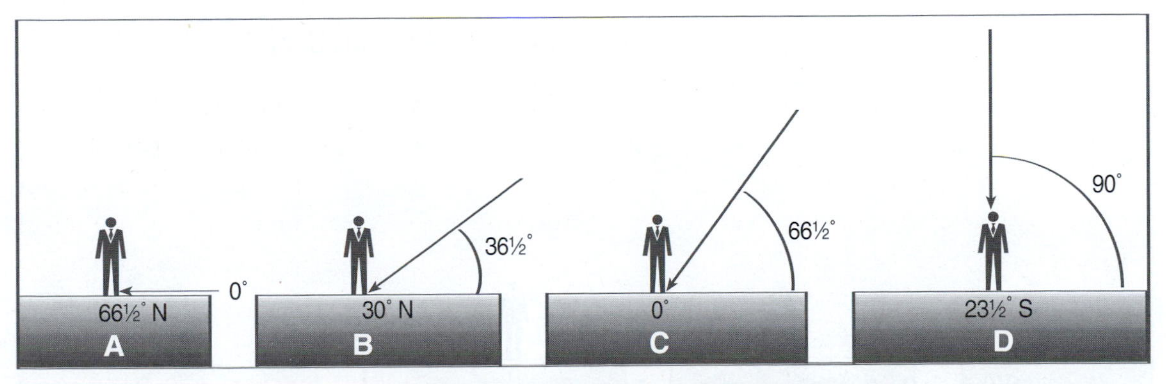

Figure 2-4. Profile view at the earth's surface, December 22.

1. *The earth–sun orientation will change throughout the year as the earth revolves around the sun. Using Figures 2-3 and 2-4 as models, sketch two similar diagrams for each date given below. (First draw the earth's axis and equator, then include all four locations on the "sunny side" of the globe.)*

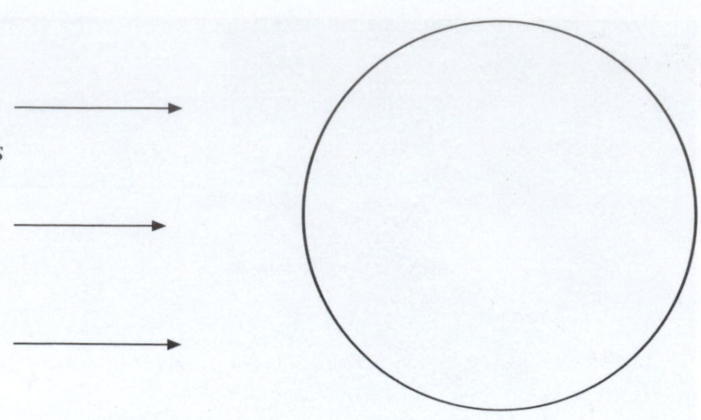

Sun's rays striking the earth on June 21

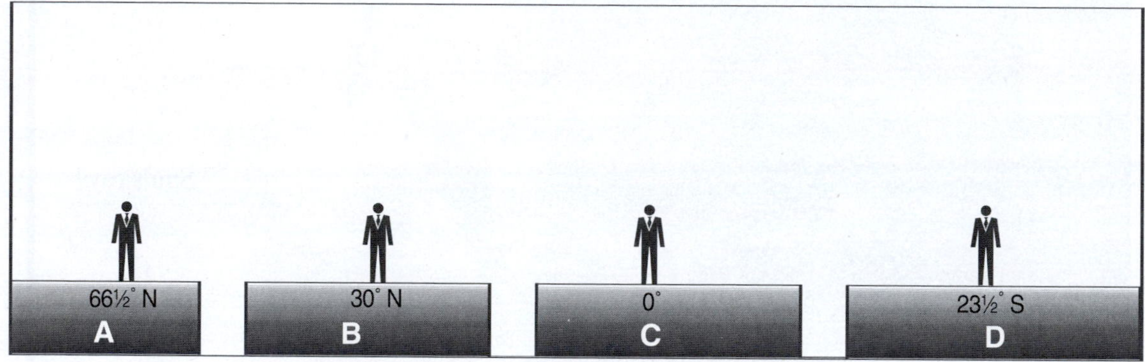

Profile view at the earth's surface: solar noon, June 21

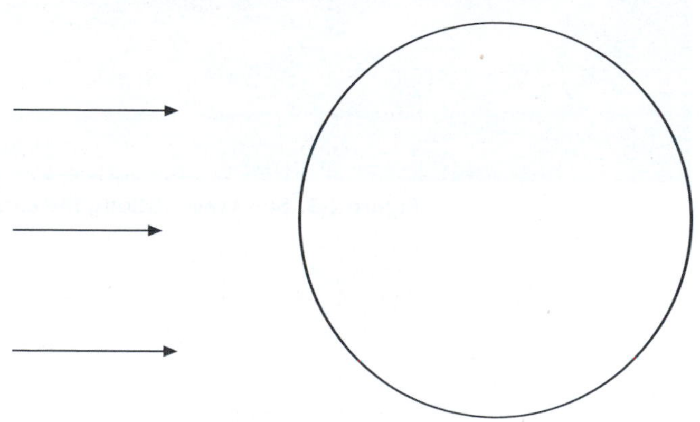

Sun's rays striking the earth on March 21

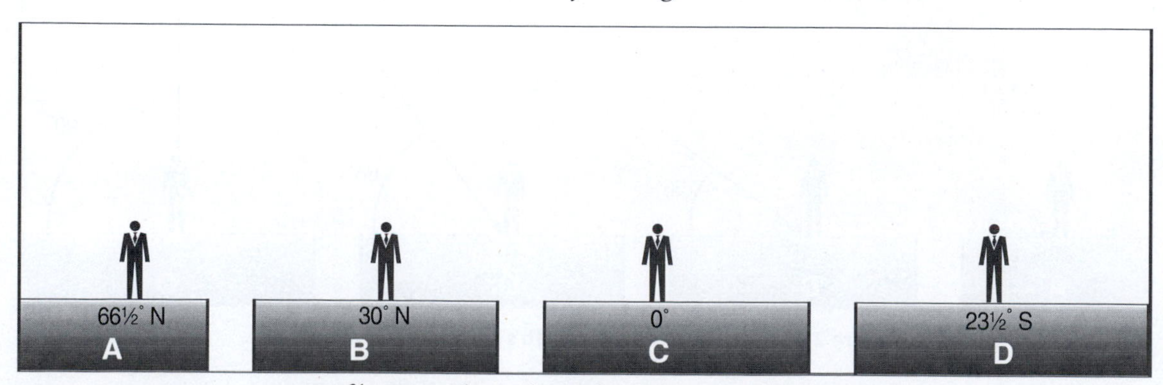

Profile view at the earth's surface: solar noon, March 21

Your diagrams should show that sun angle varies with season and location. Since such variability greatly influences weather patterns, it is useful to be able to calculate the noon sun angle for a given latitude. We must first define a few terms:

- *solar declination*—the latitude at which the sun is directly overhead at solar noon
- *zenith angle*—the angle between a point directly overhead and the sun at solar noon
- *solar elevation (sun) angle*—the angle of the sun above the horizon at solar noon

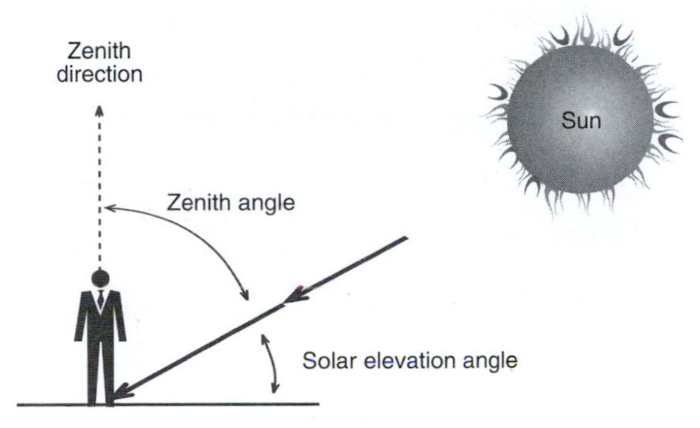

Figure 2-5

Since we can assume that the sun's rays strike the earth in straight, parallel paths, we see that the zenith angle of any location is the same as the number of degrees separating the location and the place receiving direct solar rays. Notice in Figure 2-6 that the zenith angle A at 40° N is the same as the latitude difference between 40° N and the solar declination (where the sun's rays strike directly —23½° S here).

2. *In Figure 2-6, how many latitude degrees separate the person at 40° N and the place receiving direct solar rays? ________° What is the date of this example?*

Notice that the zenith angle plus the solar elevation angle sum to 90°. Solar elevation always equals 90° minus the zenith angle.

3. *What is the solar elevation angle in Figure 2-6?*

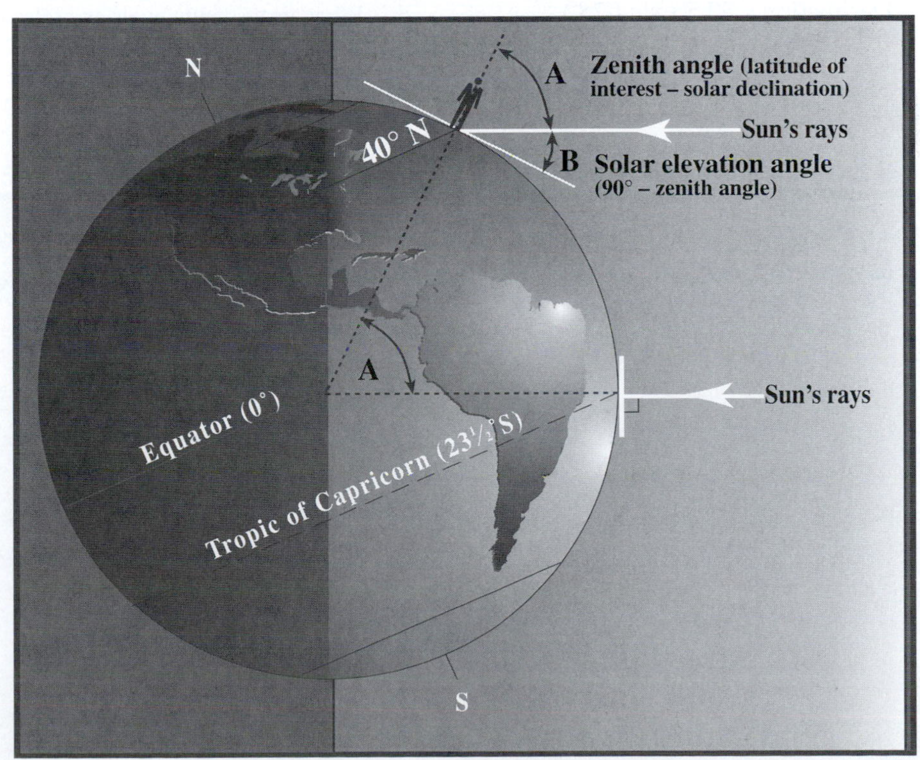

Figure 2-6. Zenith and solar elevation angles.

Figure 2-2 gives the solar declination for the solstices and equinoxes, but what about the other days of the year? You can approximate the value of the solar declination using the following formula:

$$\text{Solar declination} \approx 23.5 \bullet \sin(N)$$

where N = the number of days to the closest equinox, expressed in degrees. (By convention, N is positive between the March and September equinoxes and negative from the September to March equinoxes.)

For example, on April 20, N = 30 (number of days from the closest equinox, March 21), and

$$\text{Declination} \approx 23.5 \bullet \sin(30°)$$
$$= 23.5 \bullet (0.5) = 11.75° \text{ or } 11° 45' \text{ N}$$

On December 9, N = -78 (number of days from September 22, negative since it is between the September and March equinoxes), and

$$\text{Declination} \approx 23.5 \bullet \sin(-78°)$$
$$= -22.90° \text{ or } 22° 53' \text{ S}$$

Notice that we use negative declination values for the Southern Hemisphere and positive ones for the Northern Hemisphere—again by convention.

4. *Calculate the solar declination on:*
   a. March 21 ______________
   b. June 21 ______________
   c. September 22 ______________
   d. December 22 ______________
   e. Today's date ______________

5. *Calculate the noon sun angle for New Orleans, USA (30° N), and for Helsinki, Finland (60° N), on each of the following dates:*

	New Orleans	*Helsinki*
a. March 21	________	________
b. June 21	________	________
c. September 22	________	________
d. December 22	________	________
e. Today's date	________	________

6. *What are the zenith angle and solar elevation angle for your city today?*

Optional Exercise: Measuring Solar Elevation Angle (Sun Angle)

This procedure should be done very close to solar noon (the midpoint between sunrise and sunset).

7. *Use the following steps to measure the solar noon sun angle at your latitude:*
   a. Set a pole with a known length at a right angle to a flat surface on the ground.
   b. Measure the length of its shadow.
   c. Divide the pole length by the shadow length to calculate the tangent of the sun angle θ.

$$\theta = \frac{\text{(length of pole)}}{\text{(length of shadow)}}$$

   d. Using a calculator or table of tangent values, convert the tangent ratio into an angle.

$$\theta = \tan^{-1}\left[\frac{\text{(length of pole)}}{\text{(length of shadow)}}\right]$$

   Today's sun angle: ____________°

8. *How does this value compare with the result of your calculation in question 6? Can you account for any differences?*

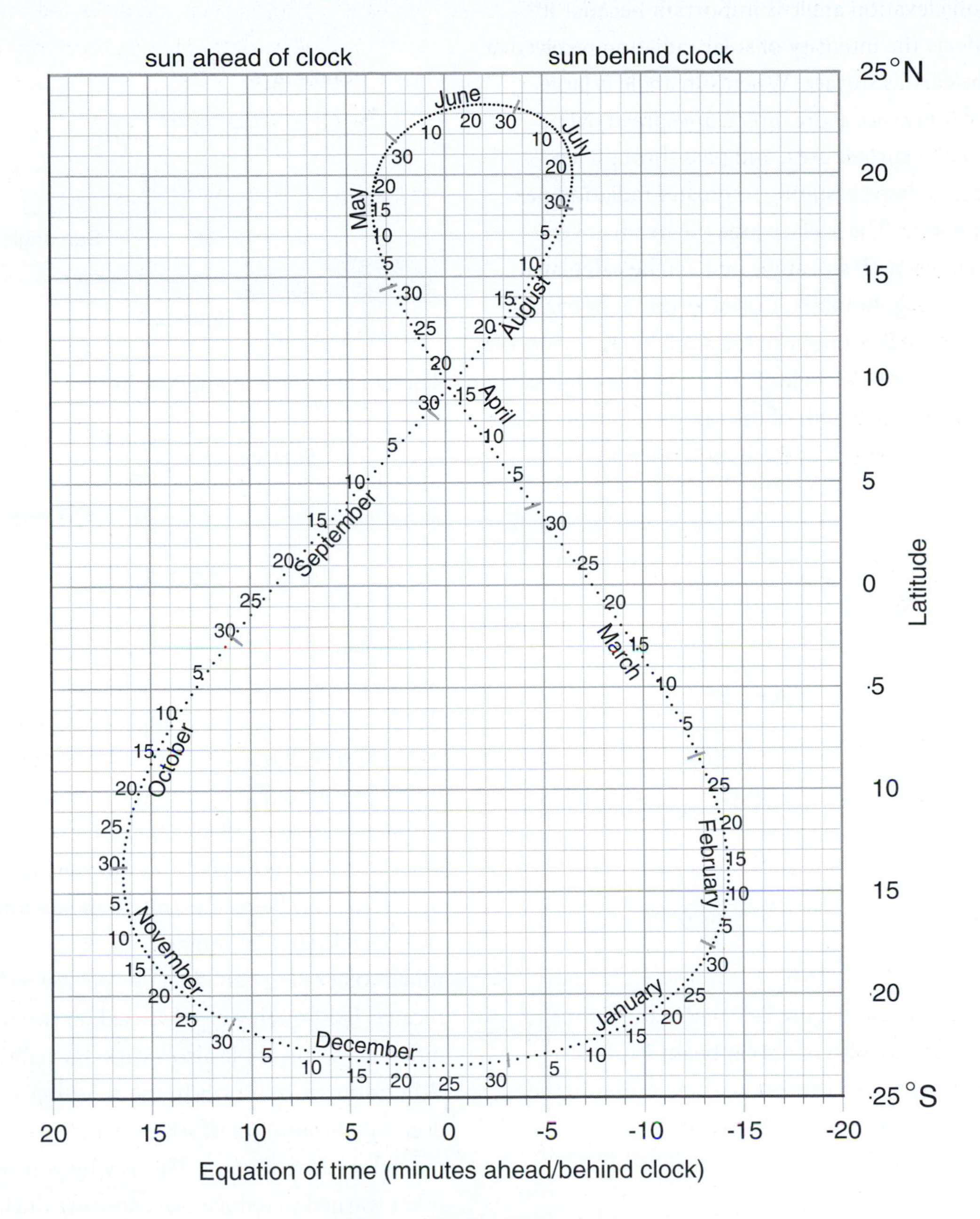

Figure 2-7. Analemma.

The analemma shown in Figure 2-7 provides —among other information—a graphical display of solar declination throughout the year. Each dot represents a given day during the year and corresponds with a latitude read on the y axis (to the right), which is the solar declination associated with that day. Notice, for example, that the declinations on the equinoxes and solstices correspond to those in Figure 2-2. Compare the declination values for the three additional dates you calculated in question 4 with those on the analemma.

Calculating Solar Intensity

Sun elevation angle is important because it affects the intensity of solar radiation received at the earth's surface. When sun angle is large, solar rays are more direct, are spread over a smaller surface area, and pass through less atmosphere, resulting in greater radiation per unit area. The surface area the beam covers changes with sun angle and can be calculated using trigonometry. Consider the right triangle in Figure 2-8. Suppose the angle α represents the solar elevation angle. The sine of this angle is equal to the length of the opposite side (*o*), divided by the length of the hypotenuse (*h*).

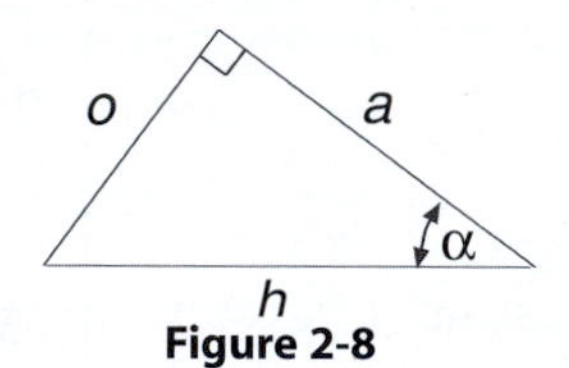

Figure 2-8

$$\sin \alpha = \frac{o}{h}$$

$$h = \frac{o}{\sin \alpha}$$

Now consider Figure 2-9, where the opposite side is the width of a solar beam 1 unit long. Since we know two of the three values in the equation, we can solve it thus:

$$\sin(\text{sun angle}) = \frac{\text{1 unit width}}{\text{surface area}}$$

$$\text{Surface area} = \frac{1}{\sin(\text{sun angle})}$$

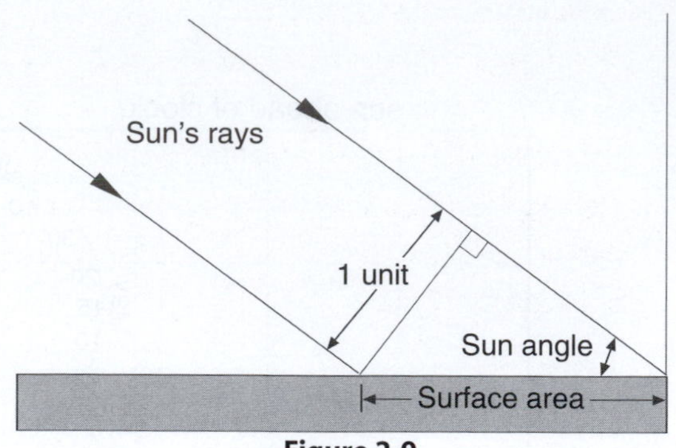

Figure 2-9

Therefore, if the sun angle is 36½°:

$$\text{Surface area} = \frac{1}{\sin(36½°)}$$

$$\text{Surface area} = \frac{1}{0.595} = 1.681$$

This means that 1 unit area of sunshine (the width of the beam shown in Figure 2-9) striking the flat earth from a 36½° angle above the horizon will spread out over 1.681 area units. Such beam spreading diminishes the maximum intensity of radiation to 59.5% (1/1.681 = 0.595) of what could be received if the sun were directly overhead. This helps illustrate how sun angle and, therefore, solar intensity determines how much solar energy an object or person will receive. (Although we've focused on seasonal differences, sun angle also varies diurnally. When you are outdoors in the morning and evening, the amount of solar energy received is much less than at noon. This is why you are often warned to reduce sun exposure during midday hours in the summer.)

9. *Following the patterns provided below, draw a simple series of sketches for solar noon on June 21 at a site at 30° N and on December 22 and June 21 for a site at 60° N. Indicate the zenith angle, solar angle, and beam spreading in each diagram. Note that you will need to change the scale of the beam for December 22 at 60° N.*

10. *Given these sketches, explain why sun angle causes seasonal temperature changes in the mid-latitudes.*

For a site at 30° N

December 22

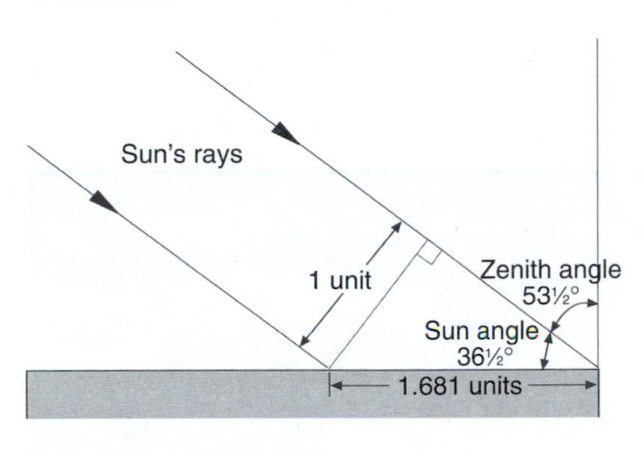

March 21

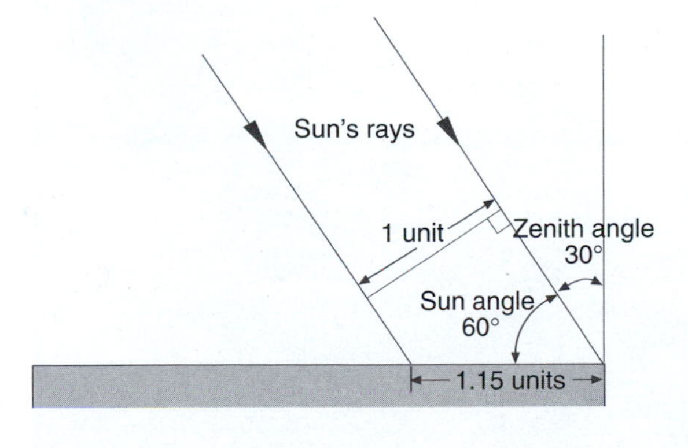

June 21

For a site at 60° N

December 22

March 21

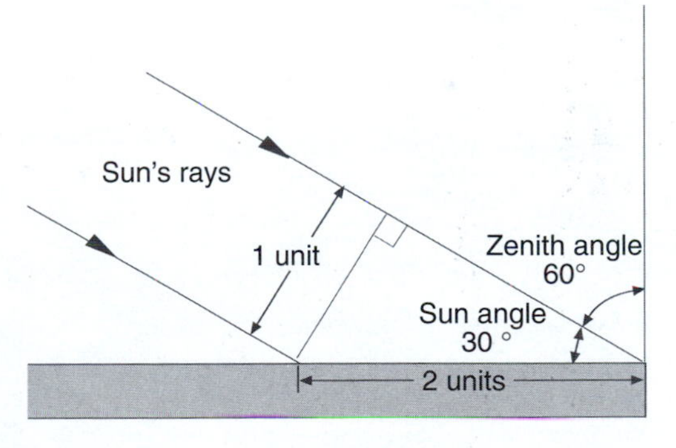

June 21

Daylight Hours

Daylight hours also have an effect on solar radiation receipt. At any given time only half of the earth is illuminated by the sun. The division between the light and the dark halves of the earth is called the *circle of illumination*. This division runs through the poles during the spring and fall equinoxes. On these dates every latitude is bisected (cut in half) and there are 12 hours of daylight and 12 hours of darkness everywhere on earth. During most of the year, however, individual lines of latitude will not be bisected but will be disproportionately divided between light and dark. Figure 2-10 illustrates this phenomenon. You can use the figure to estimate the proportion of each latitude that is illuminated during the 24-hour day.

11. *Describe the seasonal changes in daylight hours in polar regions and in tropical regions.*

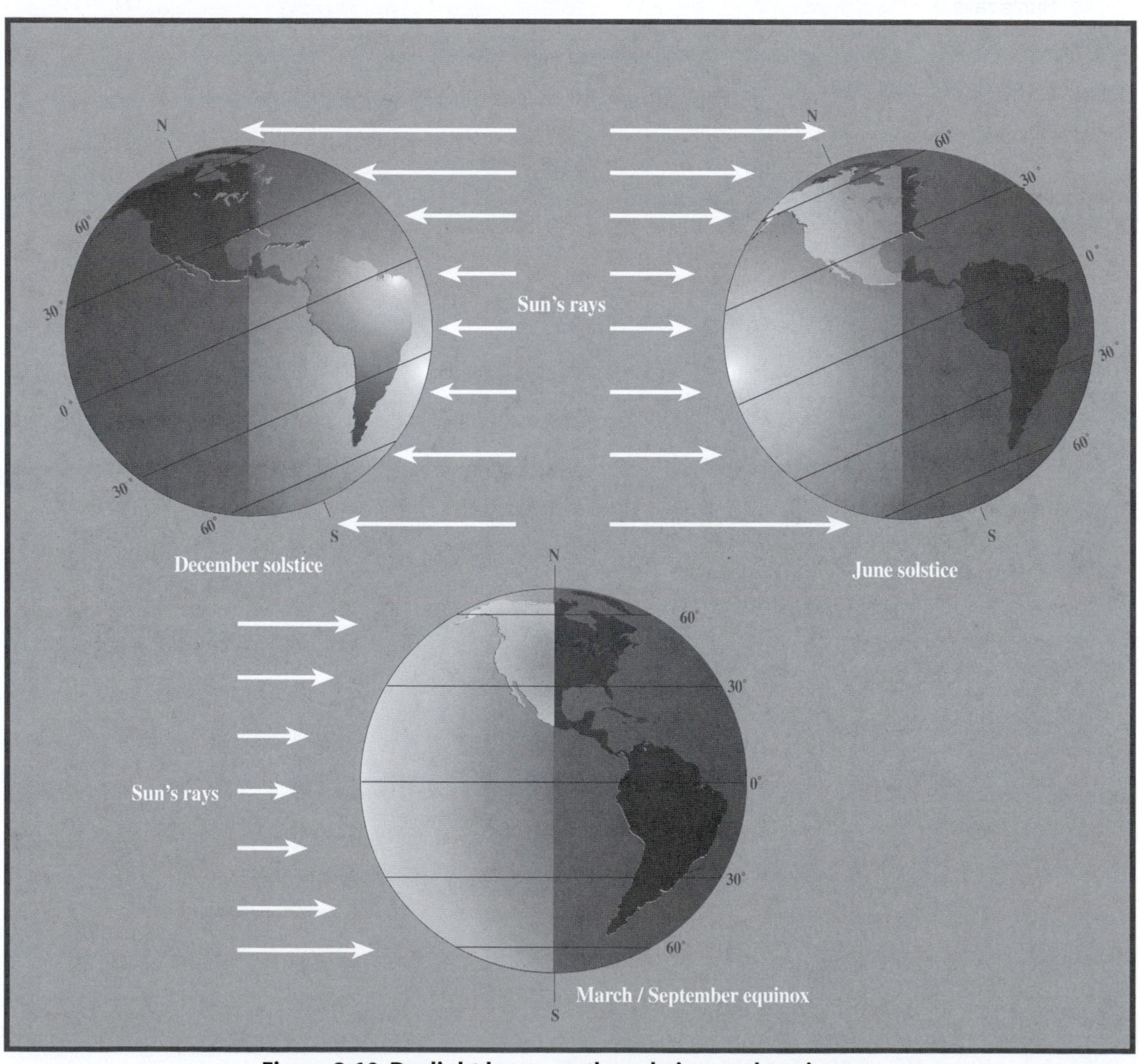

Figure 2-10. Daylight hours on the solstices and equinoxes.

Figure 2-11 shows the daylight hours during the course of the year for the equator, 30° N, and 60° N latitude.

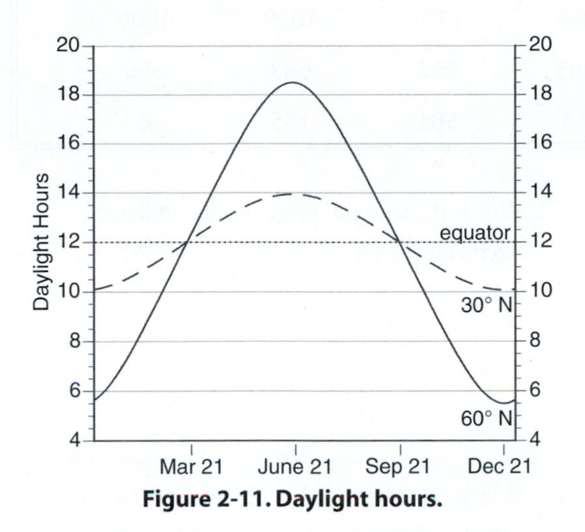

Figure 2-11. Daylight hours.

12. *Complete the table below by reading Figure 2-11 and noting the approximate number of daylight hours for each date and locations.*

	30° N	**60° N**
June solstice	______	______
Equinoxes	______	______
December solstice	______	______

13. *Which latitude experienced the greatest seasonal change in daylight hours?*

14. *Daylight hours increase or decrease incrementally from one day to the next. During what time(s) of the year is the daylight hour change greatest? When is it smallest?*

Figures 2-12 and 2-13 can be used to calculate the solar elevation angle and daylength at two latitudes (30° N and 60° N) for the solstices and equinoxes. The circles in these graphs show solar elevation angle, from 0° on the outside to 90° in the center. The angle labels on the outside of the outer circle measure the azimuth angle relative to north (i.e., 90° is east, 180° is south, etc.). Each black curve shows the course of the sun during an individual day. For example, Figure 2-12 shows that on the equinoxes at 30° N the sun rises directly in the east (90°), sets directly in the west (270°), and reaches its peak sun angle (60°) at solar noon.

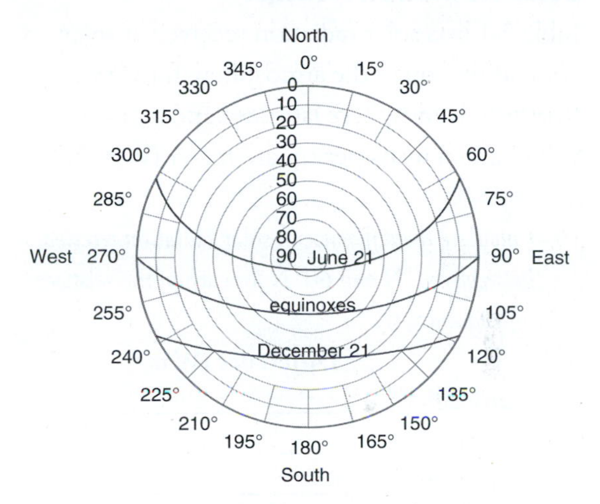

Figure 2-12. Solar angle, azimuth, and daylight hours at 30° N.

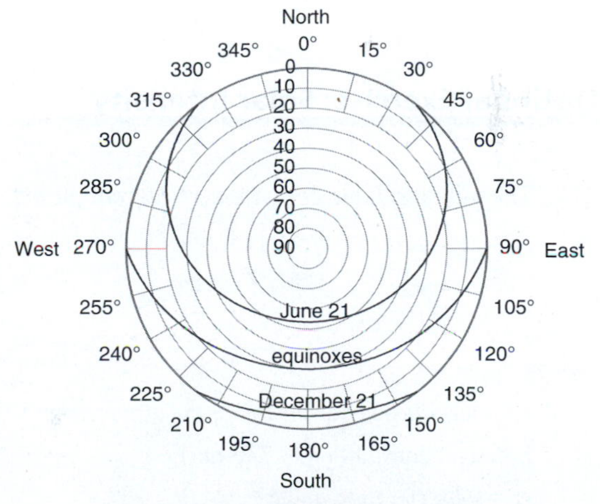

Figure 2-13. Solar angle, azimuth, and daylight hours at 60° N.

15. a. *Describe the seasonal difference in the direction in which the sun rises and sets at 30° N.*

    b. *How does it differ at 60° N?*

Table 2-1

	Equator	$23^1/_2$° N	30° N	45° N	60° N	$66^1/_2$° N
March 21	1367	1253	1183	967	683	545
June 21	1253	1367	1358	1272	1009	1000
September 22	1367	1253	1183	967	683	556
December 22	1253	932	813	501	155	0

Solar Radiation Receipt

Table 2-1 lists solar radiation received at solar noon at the top of the atmosphere for various Northern Hemisphere latitudes and dates. Values are in watts per square meter (W m^{-2}).

16. *Calculate the difference in solar radiation received between 30° N and 60° N for each of the solstices and equinoxes.*

	Difference (W m^{-2})
March 21	________________
June 21	________________
September 22	________________
December 22	________________

17. *Why is the seasonal range in solar radiation received greater at 60° N than at 30° N?*

18. *Why is the difference in solar receipt between 60° N and 30° N greater in the winter than in the summer?*

Optional Exercise: Solar Intensity

19. *The values in Table 2-1 were derived from the following equation:*

$$I = I_o \cdot \sin \alpha$$

where:

I = *solar intensity at the top of the atmosphere*
I_o = *solar constant (1367 W m^{-2})*
α = *solar elevation angle*

Calculate today's solar receipt at the top of the atmosphere at solar noon for your location.

Interactive Computer Exercise: Earth–Sun Geometry

This module allows you to see how solar angle and solar intensity change with latitude and season. The questions below encourage you to compare solar intensity at one latitude against another. You might also want to experiment to answer your own questions.

Launch the module and stop the animation on January 1. Note the settings:

Date: January 1

Solar declination: 23.1° S

	Bottom Left	**Bottom Right**
Latitude:	0° (equator)	23.5° N
Sun angle:	66.9°	43.4°
Beam spreading per unit beam:	1.087 surface units	1.455 surface units

We could calculate the percentage difference in beam spreading between the two locations as:

$$\frac{1.455 - 1.087}{1.087} = 0.34$$

Solar intensity is 34% greater at the equator than at 23½° N on January 1.

Press the play button (▸) to simulate the earth's revolution around the sun.

20. *What is the annual range of beam spreading at the equator?*

Most direct rays:

1 unit beam = ____________ surface units

Date ____________________

Least direct rays:

1 unit beam = ____________ surface units

Date ____________________

21. *What is the percentage difference in beam spreading between the highest and lowest amounts at the equator?*

22. *Use the right arrow in the bottom right diagram to change the latitude from 23½° N to a latitude further north (perhaps your latitude). What is the range of beam spreading at this new location?*

Most direct rays:

1 unit beam = ____________ surface units

Date ____________________

Least direct rays:

1 unit beam = ____________ surface units

Date ____________________

23. *What is the percentage difference in solar intensity between the highest and lowest amounts at this new latitude?*

24. *Make a general statement about the relationship between latitude and the seasonal range in beam spreading (solar intensity). How does this relationship help explain why the annual temperature range in the tropics is different from that in high latitudes?*

25. *Use the top arrows to change the date to the December solstice and record the beam spreading (surface units) at each latitude below. Then change the date to the June solstice and record beam spreading.*

	December Solstice	***June Solstice***
60° N	___________	___________
50° N	___________	___________
40° N	___________	___________
30° N	___________	___________
20° N	___________	___________

26. *What do your results reveal about seasonal contrasts in the gradient of solar intensity across the mid-latitudes?*

Review Questions

Review your June 21 and December 22 sun-angle calculations for New Orleans and Helsinki. The seasonal sun-angle difference should be the same for both stations—exactly equal to 47°, the difference between the Tropic of Cancer (23½° N) and the Tropic of Capricorn (23½° S). If this is true, then why are seasonal differences in solar intensity (Table 2-1) so much greater at Helsinki?

From the examples you've seen in this lab, describe how seasonal changes in solar intensity and daylight hours at a given place should influence its annual temperature range.

Lab 3

The Surface Energy Budget

Introduction

This lab introduces you to radiation laws and the fluxes of radiation and other energy forms at the earth's surface. You will study how seasonal, diurnal, and meteorological factors influence these energy exchanges using values measured in the real world.

Radiation Laws

As we consider the earth's surface energy budget it is important to review two radiation laws that help us understand how temperature affects the amount and character of radiation emitted by a body. First, the *Stefan-Boltzmann law* states that the total energy emitted by a body increases with temperature. Mathematically:

$$E = \sigma T^4$$

where E = radiation emitted (W m^{-2})
σ = the Stefan-Boltzmann constant ($5.67 \cdot 10^{-8}$ W m^{-2} K^{-4})
T = temperature (K)

Consider the earth, which has a blackbody temperature of 255 K. How much radiation is emitted by its surface?

$$E = (5.67 \cdot 10^{-8}\ \text{W m}^{-2}\ \text{K}^{-4}) \cdot (255\ \text{K})^4$$
$$E = (5.67 \cdot 10^{-8}\ \text{W m}^{-2}\ \cancel{\text{K}^{-4}}) \cdot (4.228 \cdot 10^{9}\ \cancel{\text{K}^{4}})$$
$$E = 240\ \text{W m}^{-2}$$

Note that the earth's average surface temperature is 288 K (15°C). This is 33 K warmer than the emission temperature because the atmosphere effectively absorbs terrestrial radiation, warms, and therefore radiates increasing amounts of energy, some of which is directed back toward the surface.

1. *a. The sun has an average surface temperature of 6000 K. How much radiation is emitted from this surface?*

   b. How much radiation is emitted from the earth's surface at 300 K?

2. *a. In question 1, how many times warmer is the sun than the earth?*

$$\left(\frac{\text{Solar temperature}}{\text{Earth temperature}}\right)$$

b. What is this number raised to the fourth power?

c. Does your result approximately equal the ratio of solar/earth emitted radiation?

Second, *Wien's law* shows that the wavelength of radiative energy depends on temperature. Specifically, it states that the wavelength of peak emission is inversely proportional to temperature:

$$\lambda_{max} = \frac{C}{T}$$

where λ_{max} = the wavelength of maximum emission (micrometers, μm)
C = Wien's constant (2898 μm K)
T = temperature (K)

3. *Calculate the wavelength of maximum emission for the earth and the sun using the emission temperatures above.*

Your calculations above explain why we often refer to solar radiation as "shortwave" and terrestrial radiation as "long-wave" radiation.

4. *In what portion of the electromagnetic spectrum (see Table 3-1) is the peak emission from the sun?*

Table 3-1. Electromagnetic Energy

Type of Energy	Wavelength (μm)
gamma	<0.0001
X-ray	0.0001 to 0.01
ultraviolet (UV)	0.01 to 0.4
visible	0.4 to 0.7
near infrared (NIR)	0.7 to 4.0
thermal infrared	4 to 100
microwave	100 to 1,000,000 (1 m)
radio	>1,000,000 (1 m)

Radiation Fluxes

From the perspective of the surface, the radiation budget is made up of the four fluxes shown in Figure 3-1.

Incoming shortwave radiation (SW↓) is the amount of global solar radiation received at the surface. It depends on location, time of year, time of day, cloudiness, and other atmospheric conditions.

Some of the solar radiation reaching the surface is reflected (SW↑) and must be subtracted from the total incoming solar radiation. This outgoing shortwave radiation depends on surface albedo,which is influenced by characteristics of the earth's surface such as color and texture. Obviously shortwave radiation fluxes are relevant only between sunrise and sunset.

We must also consider the long-wave radiation emitted by the earth's surface (LW↑) and long-wave radiation received at the earth's surface (LW↓) by the atmosphere. As you learned, outgoing long-wave radiation emitted by the earth depends on the surface temperature. Incoming long-wave radiation received by the earth depends on the state of the atmosphere. Unlike the shortwave fluxes, long-wave fluxes occur throughout the entire day.

Figures 3-2 and 3-3 on the following page show shortwave and long-wave radiation measured at the earth's surface at Maun, Botswana* (20° S latitude). Examine the four radiation fluxes for February 1–3, 2000, and August 1–3, 2000.

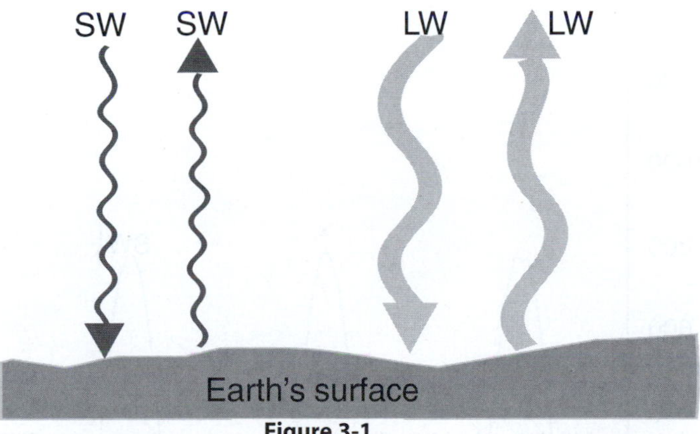

Figure 3-1

SW ↓ Incoming shortwave radiation measured at the earth's surface.
SW ↑ Shortwave radiation reflected by the earth's surface.
LW ↓ Incoming long-wave radiation emitted by the atmosphere.
LW ↑ Outgoing long-wave radiation emitted by the earth's surface.

*Data from Lloyd, J., O. Kolle, E. Veenendaal, A. Arneth, and P. Wolski. 2004. SAFARI 2000 Meteorological and Flux Tower Measurements in Maun, Botswana, 2000. Data set. Available on-line [http://daac.ornl.gov/] from Oak Ridge National Laboratory Distributed Active Archive Center, Oak Ridge, Tennessee, U.S.A. doi:10.3334/ORNL-DAAC/760.

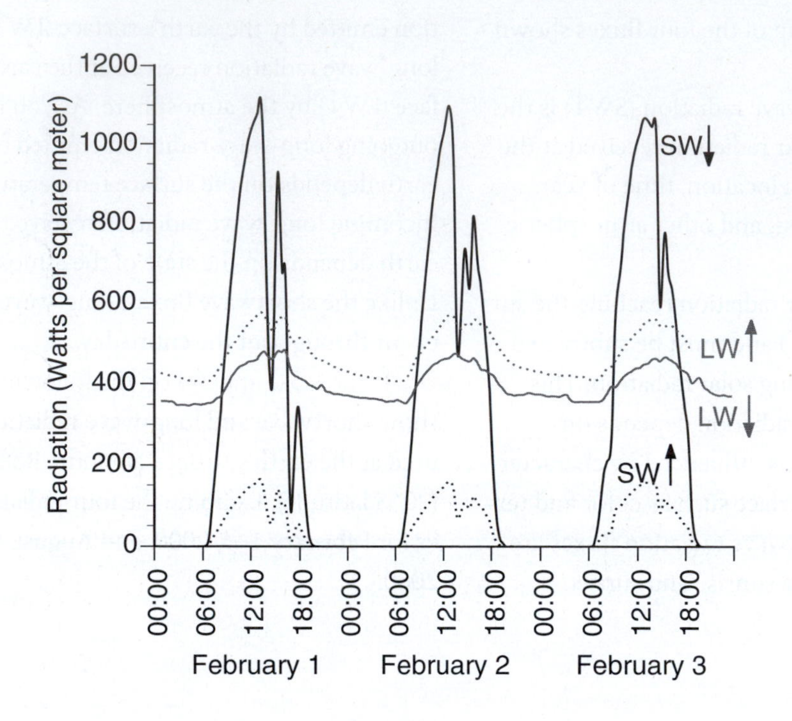

Figure 3-2. Hourly radiation fluxes—Maun, Botswana, February 1–3, 2000.

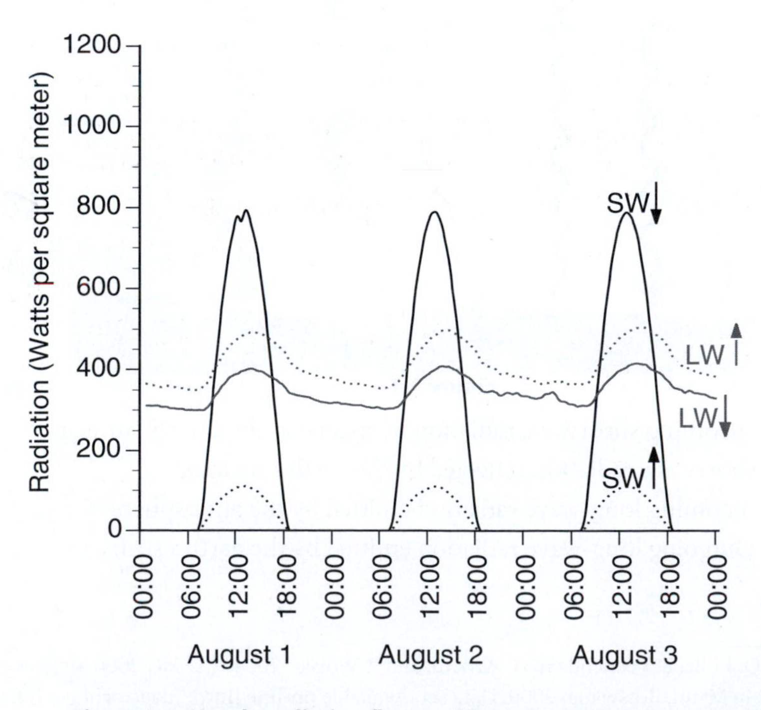

Figure 3-3. Hourly radiation fluxes—Maun, Botswana, August 1–3, 2000.

5. *Why is incoming short-wave radiation higher in early February than it is in early August at this location?*

6. *The solar noon solar radiation values at the top of the atmosphere above Maun are 1366 Wm^{-2} on February 2 and 1074 Wm^{-2} on August 2. Why is solar receipt at the surface lower?*

7. *What do the incoming short-wave radiation curves suggest about cloudiness on February 1–3 and August 1–3?*

8. *At 12:30 PM on August 2, the surface receives 789 Wm^{-2} and reflects 110 Wm^{-2}. What is the surface albedo?*

$$\left(\text{Albedo} = \frac{SW\uparrow}{SW\downarrow} \bullet 100\%\right)$$

9. *Why does outgoing long-wave radiation from the earth's surface exceed the long-wave radiation received from the atmosphere in these examples?*

10. *Look closely at the curves showing long-wave radiation emission from the earth's surface on the six dates. Explain what accounts for the general pattern you see each day, and what causes the subtle differences in the magnitude and timing of peak long-wave radiation emitted by the earth's surface between February 1-3 and August 1-3?*

Net Surface Radiation and Other Energy Fluxes

The net surface radiation is calculated as the sum of incoming shortwave (SW↓) and long-wave (LW↓) radiation minus the total outgoing shortwave (SW↑) and long-wave (LW↑) radiation:

$$\text{Net surface radiation} = SW\downarrow - SW\uparrow + LW\downarrow - LW\uparrow$$

Figures 3-4 and 3-5 show the net surface radiation at Maun on March 19, 2000, and September 24, 2000, respectively. Note that these dates are near the equinoxes, and solar intensity at Maun is about the same (the solar noon zenith angle is approximately 20°, resulting in approximately 1285 Wm^{-2} of solar radiation at the top of the atmosphere).

The energy from surplus radiation at the surface can be used to heat the atmosphere via conduction or convection (sensible heat flux), heat the ground (soil heat flux), or evaporate moisture (latent heat flux). In Figures 3-4 and 3-5, positive values for each of these fluxes denote an energy gain by the earth's surface; negative values denote an energy loss by the earth's surface.

The proportion of each of these fluxes depends, in part, on moisture availability at the surface. In Maun, moisture availability changes seasonally because of a marked rainy season. Figure 3-6 shows Maun's average monthly maximum and minimum temperatures (°C) and average monthly precipitation (cm).

11. *How and why does the net surface radiation pattern differ between the two dates during daylight hours?*

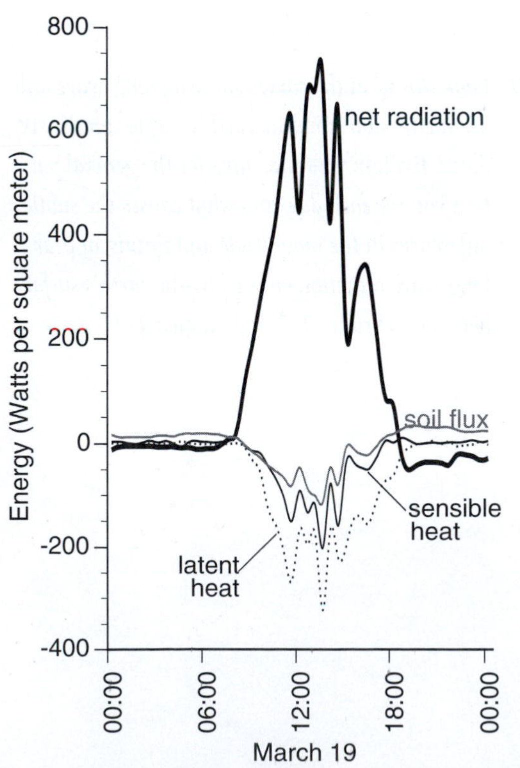

Figure 3-4. Net surface radiation—Maun, Botswana, March 19, 2000.

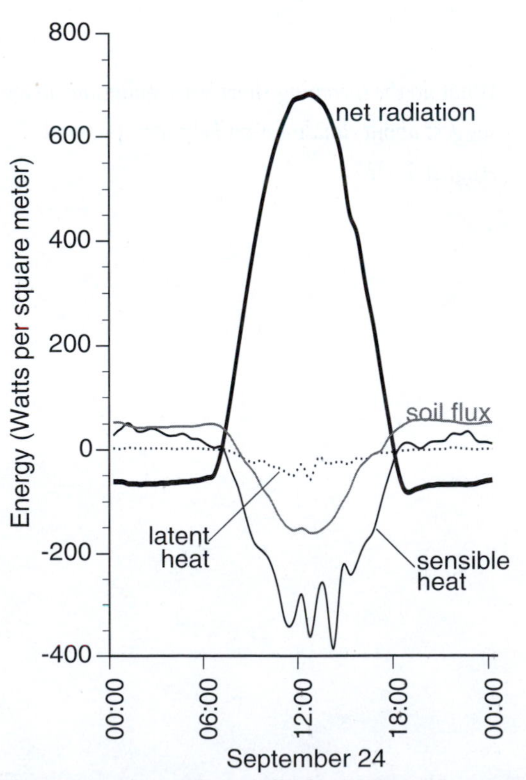

Figure 3-5. Net surface radiation—Maun, Botswana, September 24, 2000.

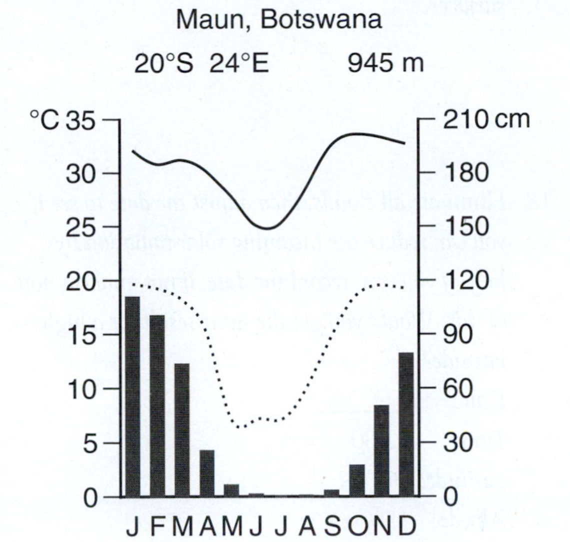

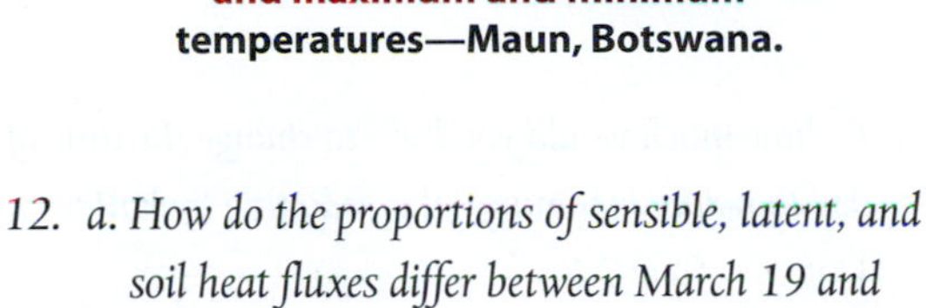

Figure 3-6. Average monthly precipitation and maximum and minimum temperatures—Maun, Botswana.

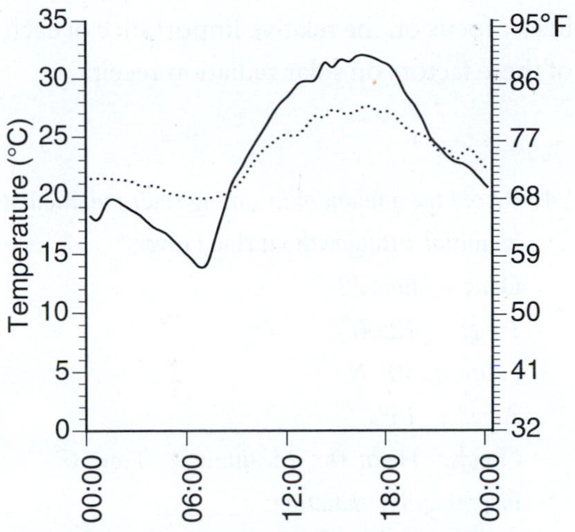

Figure 3-7. Diurnal air temperatures—Maun, Botswana, March 19 and September 24, 2000.

12. *a. How do the proportions of sensible, latent, and soil heat fluxes differ between March 19 and September 24?*

    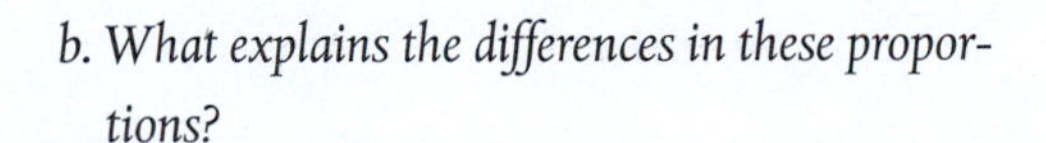

    b. What explains the differences in these proportions?

13. *a. Figure 3-7 shows the diurnal air temperature 1.5 meters above the surface on March 19 and September 24. Based on the distribution of sensible, latent, and soil heat fluxes, match each curve to the appropriate date:*

    Solid curve date ____________________

    Dotted curve date ____________________

    b. Explain your selection.

Interactive Computer Exercise: Shortwave Radiation

This exercise allows you to examine how cloud cover, latitude, season, and time of day affect incoming shortwave radiation. The questions below focus on the relative importance of each of these factors on solar radiation receipt.

14. *Record the amount of incoming solar radiation at an initial setting without cloud cover.*
    Date: June 22
    Time: 12:00
    Latitude: 30° N
    Albedo: 15%
    Clouds: High: 0 Medium: 0 Low: 0
    Incoming SW radiation: _______

15. *Compare this value with the amount of solar radiation at the top of the atmosphere (Table 2-1). What do you think accounts for the reduction in solar radiation striking the earth's surface?*

16. *Add some combination of clouds to reduce incoming solar radiation by approximately half. Record your results.*
    Date: June 22
    Time: 12:00
    Latitude: 30° N
    Albedo: 15%
    Clouds: High: ____ Medium: ____ Low: ____
    Incoming SW radiation: _______

17. *At which level do the clouds seem to have the greatest effect on the solar radiation received at the surface?*

18. *Eliminate all clouds, then adjust the date to see if you can reduce the incoming solar radiation by half. If you can, record the date; if not, make a note of this. Would you get the same result for a high latitude?*
    Date: _______
    Time: 12:00
    Latitude: 30° N
    Albedo: 15%
    Clouds: High: 0 Medium: 0 Low: 0
    Incoming SW radiation: _______

19. *By how much would you have to change the time of day to reduce incoming solar radiation by half?*
    Date: June 22
    Time: _______
    Latitude: 30° N
    Albedo: 15%
    Clouds: High: 0 Medium: 0 Low: 0
    Incoming SW radiation: _______

20. *How does albedo influence the net radiation (incoming shortwave radiation minus reflected shortwave radiation) at the earth's surface?*

Interactive Computer Exercise: Long-Wave Radiation

This exercise will let you explore the relationship between the earth's surface temperature and its emission of long-wave radiation. It also includes a simplified treatment of atmospheric long-wave emission back toward the earth's surface.

21. *How does long-wave radiation emission from the earth's surface change with increasing temperature?*

22. *Note that the difference in surface emission between -10 °C and -5 °C is not the same as the difference in surface emission between 35 °C and 40 °C. Use one of the radiation laws to explain why this is the case.*

23. *How does simulated incoming long-wave radiation change with increasing cloudiness? How does the amount of incoming long-wave radiation depend on cloud height?*

The exercise provides a simplified treatment of long-wave emission from the atmosphere that is dependent only on cloud amount. In reality, atmospheric temperature will determine how much radiation is radiated back toward the earth's surface.

24. *What factors besides clouds could alter atmospheric temperature and, therefore, atmospheric emission of radiation?*

25. *Cloudy nights do not cool as fast as clear nights. How does this module illustrate the processes causing this phenomenon? How does it limit consideration of this phenomenon (i.e., is there a variable that should change in response to a radiation flux but does not)?*

Review Questions

Which varies more seasonally, incoming short-wave (solar) radiation or outgoing long-wave (terrestrial) radiation? Which varies more diurnally? Explain your answers.

How would predictions of cloud cover influence a forecast of tomorrow's high and low temperatures?

Lab 4

The Global Energy Budget

Introduction

In Lab 3 we examined energy fluxes at a local scale; here we will consider the global energy budget, especially the differences in energy across the globe. It is the differences in energy from place to place that drive most atmospheric processes.

The Global Energy Budget and Temperature

As you know, temperature varies by location and season. You have already studied the importance of earth-sun geometry and solar intensity. Let's apply those concepts to the global radiation budget and its effect on temperature.

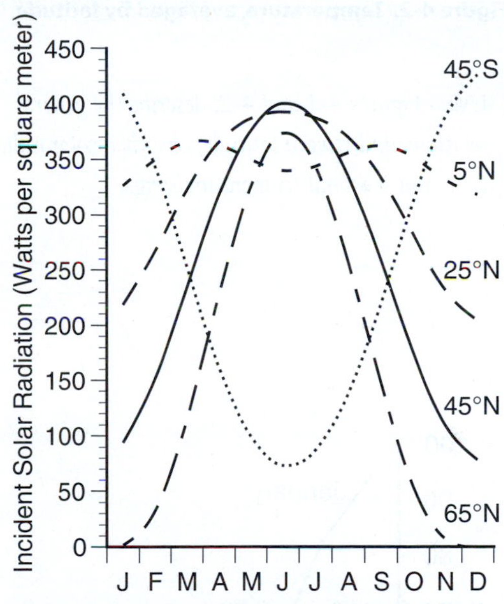

Figure 4-1. Solar radiation striking the earth's surface, averaged by latitude.

1. *At 5° N latitude there are two peaks of incident solar radiation (that is, solar radiation striking earth's surface) during the year. Why does this location differ from other latitudes that have only one peak?*

2. *The earth in its elliptical orbit is closest to the sun on January 3 (the* perihelion*) and farthest from the sun on July 3 (the* aphelion*). What evidence is there in Figure 4-1 that this affects incident solar radiation?*

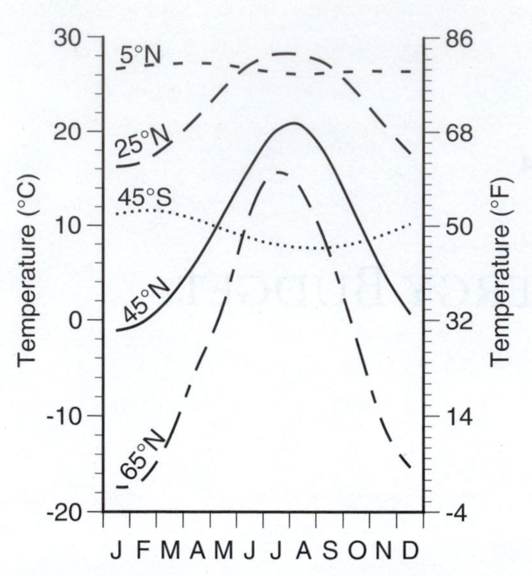

Figure 4-2. Temperature, averaged by latitude.

3. *Using Figures 4-1 and 4-2, describe the general relationship between latitude, absorbed solar radiation, and seasonal temperature range.*

Albedo

Albedo, or reflectivity, at the earth's surface is a function of sun angle and of the color and texture of the surface. Albedo values range from 0% (no reflection) to 100% (complete reflection). Examine Figures 4-3 and 4-4.

4. *Why is there little seasonal difference in albedo in the tropics and across most of the Southern Hemisphere, but greater seasonal difference in albedo north of 40° N?*

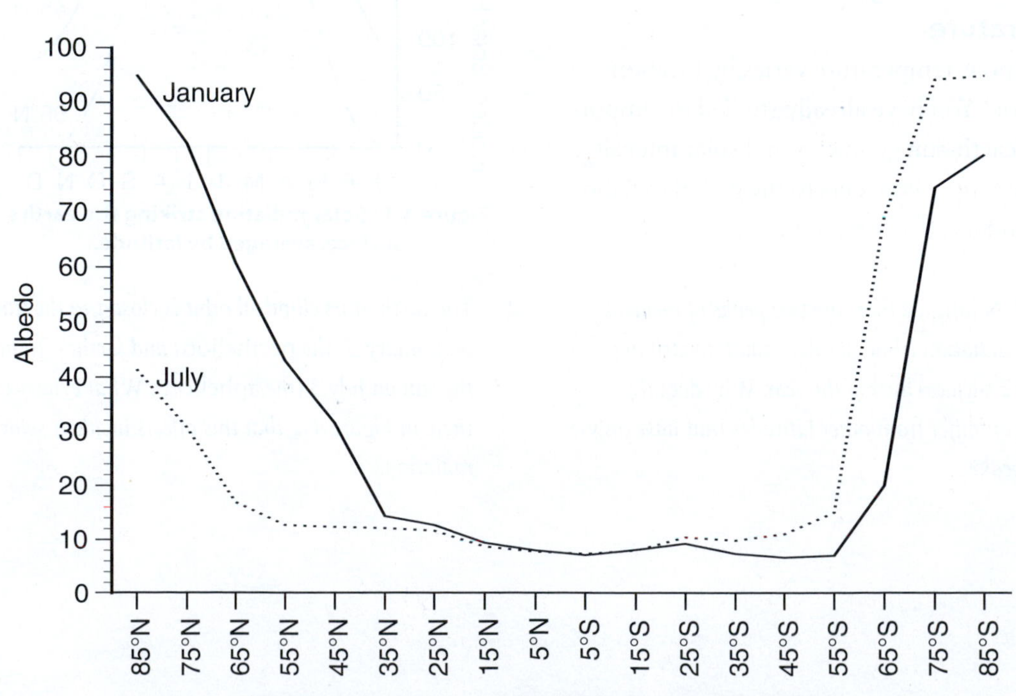

Figure 4-3. January and July albedo, averaged by latitude.

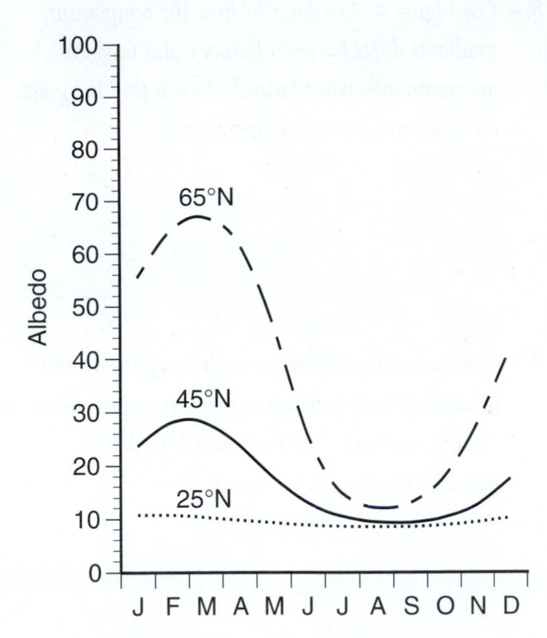

Figure 4-4. Seasonal variations in albedo at 25° N, 45° N, and 65° N.

Net Radiation and Circulation

Many of the earth's atmospheric processes result from differences in energy from one place to another. For example, global-scale atmospheric and oceanic circulation are the result of an energy surplus in tropical regions and an energy deficit in polar regions. Both sensible and latent heat are transferred by winds. Warm ocean currents, often found on the east coasts of continents, also transport warm water from tropical regions to higher latitudes in both hemispheres. The magnitude of atmospheric and circulation depends on the magnitude of differences in energy from one place to another.

5. *Examine Figure 4-5 below. When averaged annually, which latitudes have a net radiation surplus and which have a deficit?*

6. *What does Figure 4-5 say about the relationship between poleward heat transport and the gradient of net radiation from one latitude to another? (This gradient, or rate of change, can be viewed as the steepness of the net radiation curve.)*

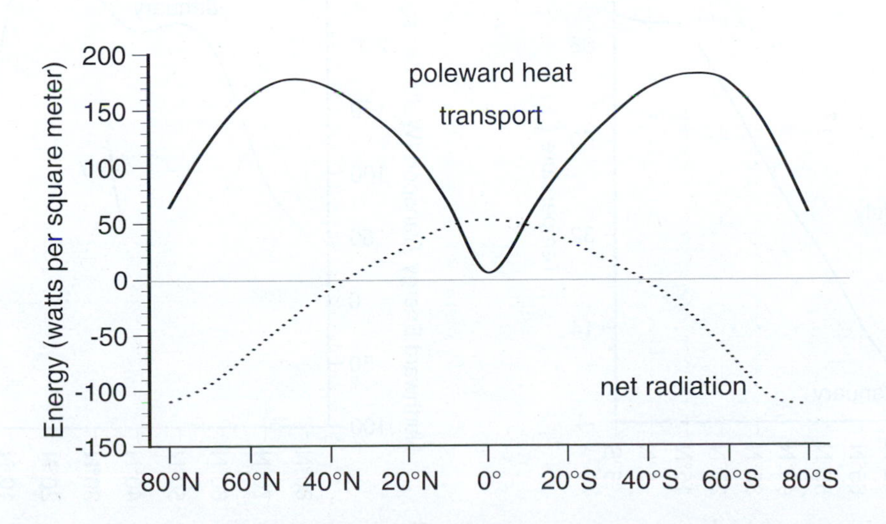

Figure 4-5. Poleward heat transport and net radiation, averaged annually and by latitude.

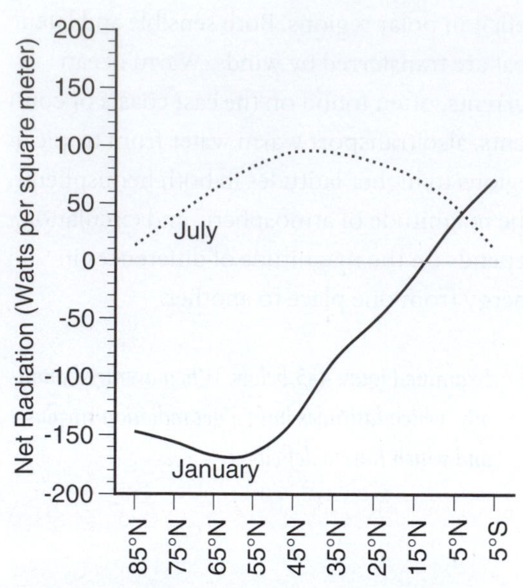

Figure 4-6. January and July, net radiation in the Northern Hemisphere, averaged by latitude.

7. *Using Figure 4-6, describe how the areas of net radiation surplus and deficit, and the latitudinal gradient of net radiation differ between January and July.*

8. *Use Figure 4-7 to describe how the temperature gradients differ between January and July. At approximately what latitude do you find the greatest range in seasonal temperatures?*

9. *Summarize the difference in the magnitude and location of peak poleward energy transport between January and July seen in Figure 4-8. What accounts for these differences?*

10. *How do the latitudinal gradients of net radiation and of temperature (Figures 4-6 and 4-7) help to explain the magnitude and location of peak poleward energy transport (Figure 4-8)?*

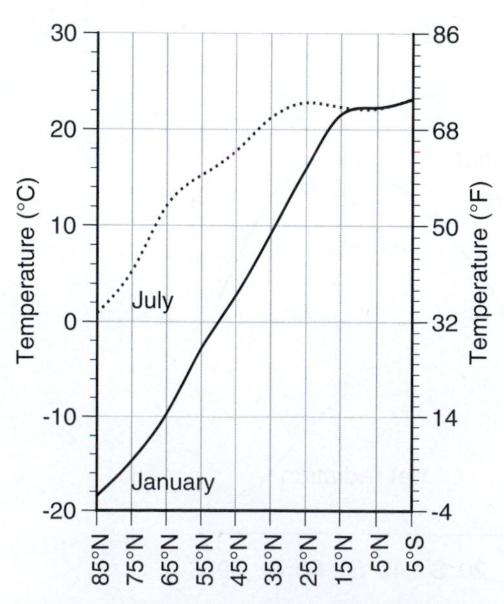

Figure 4-7. January and July temperatures in the Northern Hemisphere, averaged by latitude.

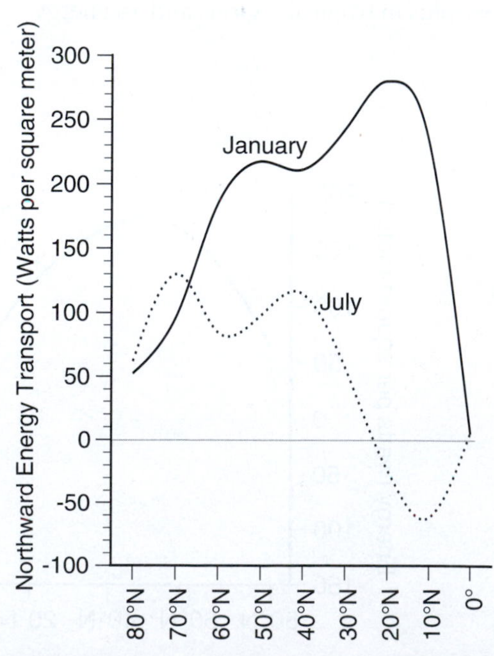

Figure 4-8. January and July energy transport in the Northern Hemisphere, averaged by latitude.

Other Factors Influencing Temperature

Water has a high specific heat, meaning that it takes larger amounts of energy to increase its temperature compared to the temperature of land surfaces. Once warmed however, water retains heat longer than most land surfaces do. So, the air temperature near a large water body will warm and cool more slowly than at a location distant from water bodies, moderating the daily and seasonal range of air temperature.

11. *Why is the seasonal temperature range greater at 45° N than at 45° S?*

In the mid-latitudes, weather patterns often move from west to east. Therefore, locations on the west coast of continents experience the moderating effect of oceans more than east-coast locations at the same latitude.

12. *Note the geographic locations of Durban, South Africa (30° S), Port Nolloth, South Africa (29° S), Punta Galera, Chile (40° S), and Viedma, Argentina (41° S), and match each city to the appropriate monthly temperature plot.*

A. ______________ B. ______________

C. ______________ D. ______________

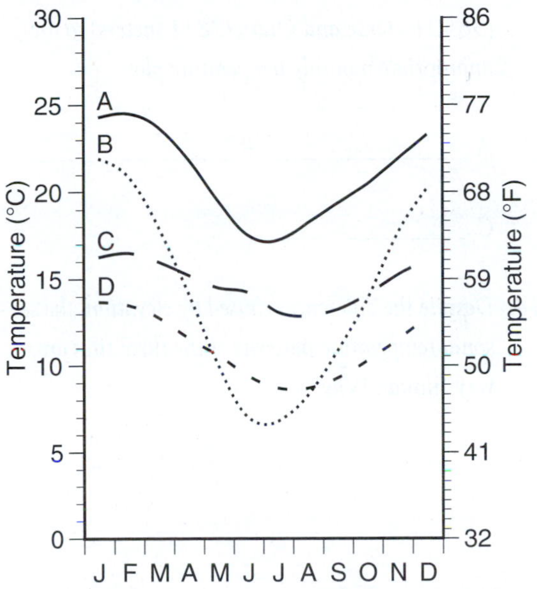

Figure 4-10. Average monthly temperatures for Durban, Port Nolloth, Punta Galera, and Viedma.

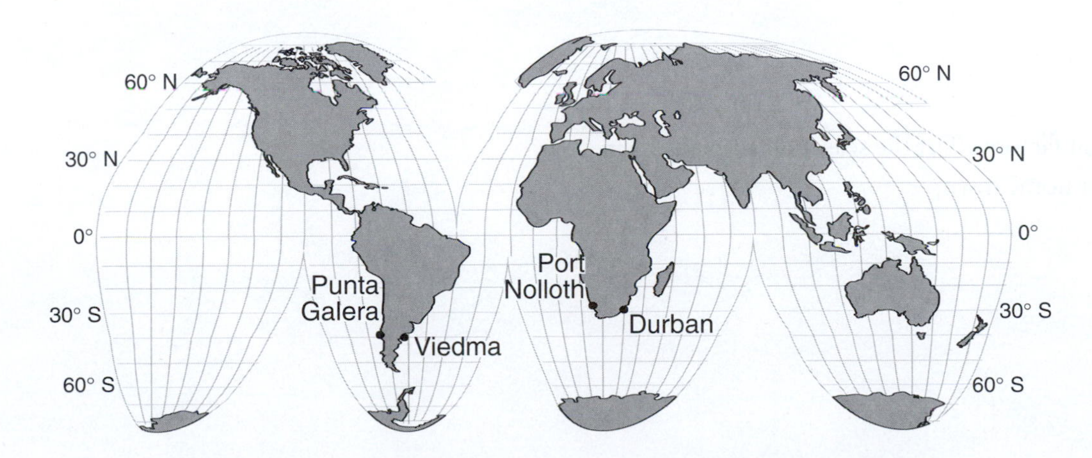

Figure 4-9. Locations of Durban, Port Nolloth, Punta Galera, and Viedma.

Elevation affects temperature as well. As we learned in Lab 1, air density decreases with elevation. Figure 4-11 shows the average monthly temperatures for three sites in Ecuador: Bahía de Caráquez (a coastal resort); Cotopaxi (the second highest peak in the country), and Quito (the capital). Ecuador, as the name suggests, straddles the equator, and all three sites lie very close to it.

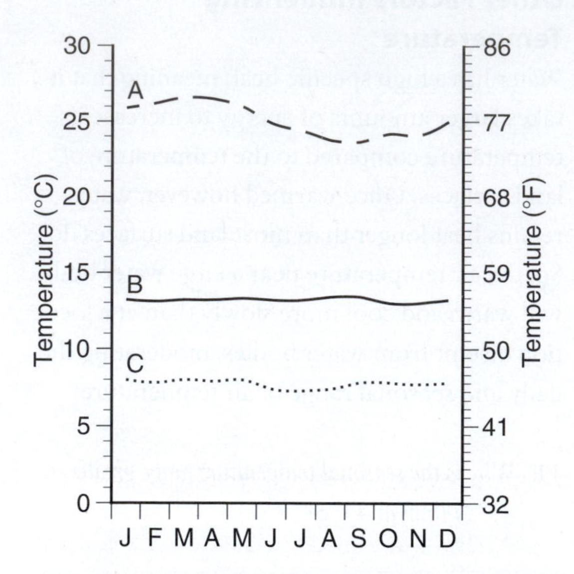

Figure 4-11. Average monthly temperatures for Bahía de Caráquez, Cotopaxi, and Quito.

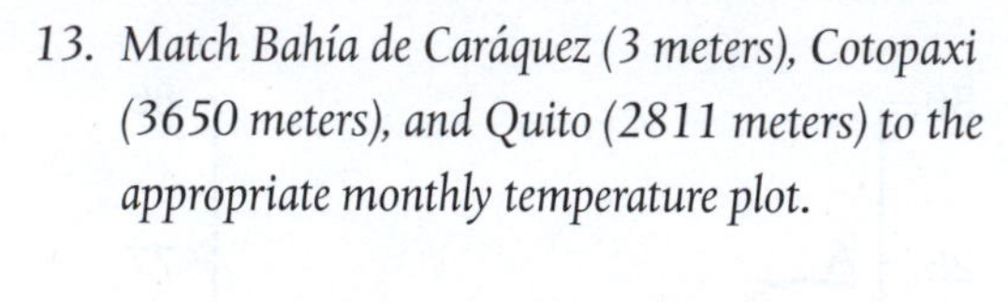

13. *Match Bahía de Caráquez (3 meters), Cotopaxi (3650 meters), and Quito (2811 meters) to the appropriate monthly temperature plot.*

    A.______________ B. ______________

    C. ______________

14. *Despite the differences caused by elevation, the seasonal temperature patterns of the three sites are very similar. Why?*

Review Questions

What determines the location and magnitude of poleward energy transport?

What factors affect the seasonal temperature in your home town?

Lab 5

Atmospheric Moisture

Materials Needed

- calculator
- ruler

Introduction

What is humidity and how is it important to atmospheric processes? This lab shows how we measure atmospheric moisture and uses two experiments to illustrate its role in energy transfer.

Latent Heat

Water is a unique substance because it occurs in three phases (solid, liquid, and vapor) at temperatures and pressures that commonly occur on earth. The phase change of water is what makes the weather interesting, variable, and often unpredictable. When it changes phase, water either releases or consumes energy—a process called *latent heat transfer.* The following two experiments illustrate the concept of latent heat. If you do not have the necessary equipment to conduct these experiments, please read through them anyway, carefully review the section on latent heat in your textbook, and answer the general questions (10–13).

Optional Exercise: Energy Transfer Experiments

Experiment I

Procedures

A. Fill a beaker with 300 milliliters (ml) of water and heat it to a boil. While the water is heating complete step B

B. Fill another beaker about ¾ full of ice and the remainder with water. Stir for 1 minute and record the temperature.
Ice-water temperature = ____________°C

C. Put 50 ml of the heated water and a thermometer in a beaker. When the temperature of the water has dropped to 80°C, add to it 50 ml of the water from the ice-water mixture in step B. Stir and record the final temperature.
Final temperature = ___________°C

Calculations

Note: A *calorie* (cal) is the energy required to raise the temperature of one gram of water one degree Celsius. One milliliter of water weighs 1 gram (g) since the density of water is 1 g/cm^3, and 1 ml = 1 cm^3.

1. *How many calories of heat were lost by the 50 ml of hot water as it cooled from 80°C to the final temperature? ___________ calories*

2. *How many calories were gained by the cold water? ___________ calories*

3. *The energy lost by the hot water should equal the amount gained by the cold water. If this was not true in your example, which steps in the experiment could have contributed to the difference?*

EXPERIMENT II

Procedures (*Steps A and B should be done concurrently*)

A. Chill an empty beaker. Weigh it and add 25 g of ice. Then, pour water from the ice-water beaker (from Experiment I) into the first beaker until it contains a total of 50 g of ice and water. Record the temperature of the contents.
Ice-water temperature = ___________°C

B. Pour 50 ml of boiling water into another beaker and allow it to cool to 80°C.

C. Pour the 80°C water into the ice-water mixture and stir. After all the ice has melted record the temperature.
Temperature = ___________°C

Calculations

4. *Based on the temperatures you measured, how many calories were lost by the 50 ml of hot water after it was added to the ice-water mixture? ___________ calories*

5. *How many calories were gained by the ice-water mixture as reflected by the temperature change? ___________ calories*

6. *Your experiment should show that only a portion of the energy lost by the hot water was used to heat the ice-water mixture. What is the energy difference between the two? ___________ calories*

7. *What happened to the energy lost by the hot water but not used to increase the temperature of the ice-water mixture?*

8. *How many calories were consumed to melt each gram of ice in your experiment?*
   ___________ *calories/gram*

9. *How does this value compare with the theoretical latent heat of melting (80 cal/g)? What are the possible sources of error in your experiment?*

General Questions

Experiment II highlights the energy transfer involved as water changes phase from solid to liquid. This process moves energy from a lower-energy state to a higher-energy state. Since this process requires energy, we say that energy is consumed. In this case, it is absorbed by the liquid water. As a meteorological example, consider a snowflake that melts as it falls through the atmosphere. The energy used to melt the snowflake comes from the surrounding air, and thus the air cools. Figure 5-1 shows the energy exchange during each phase change.

10. *Once water has been heated to its boiling point, additional heating will not raise its temperature. What happens to this additional energy?*

11. *Imagine two adjacent playing fields, one with a wet surface and the other with a dry surface. If you simultaneously measured air temperature above each surface on a sunny summer afternoon, where would you expect the higher temperature? Why?*

12. *Condensation is an important phase change that commonly occurs in the atmosphere. Suppose a 1-m^3 sample of air at sea-level pressure, containing 4 g of water vapor, was cooled such that all 4 g condensed. How many calories of heat would be released by this process?*

13. *Examine Figure 5-1 on the following page and fill out the chart below it with other examples of water's phase changes, noting whether energy is released or consumed, and where the energy went during the phase change.*

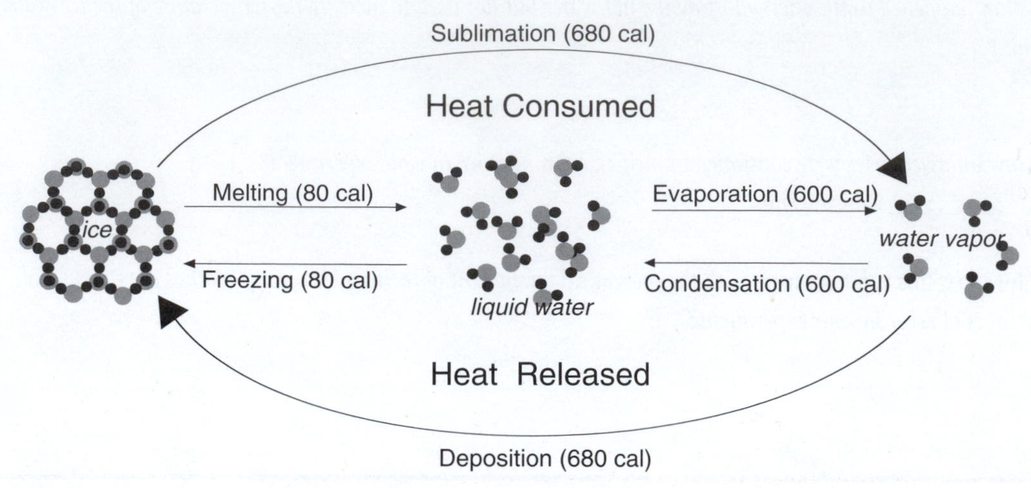

Figure 5-1. Energy transfer during phase changes.

Phase change	Energy (released or consumed)	Examples
Ice → (liquid) water	consumed; energy used to melt the ice is now contained in the water	ice cubes melting in a glass; melting of falling snowflake
Water → water vapor		
Ice → water vapor		
Water vapor → water		
Water vapor → ice		
Water → ice		

Measures of Atmospheric Moisture

The water vapor in the atmosphere comes from sources on the earth's surface such as oceans, lakes, streams, soil, and plants. The amount of water vapor present depends on the amount of energy available to change this liquid surface water to gas. As the energy supply at the earth's surface increases, liquid water molecules move faster and are more likely to leave the liquid surface and evaporate. For a given amount of energy at the earth's surface, there exists an upper limit to the amount of liquid water that can be evaporated.

There are several ways to measure atmospheric moisture at a given time and place. In one method, water vapor is considered to be like any other atmospheric gas, in that water vapor molecules exert a partial pressure proportional to their concentration in the atmosphere. This pressure is referred to as the *vapor pressure.* The "upper limit" of atmospheric vapor pressure is referred to as *saturation vapor pressure.* Since evaporation rate depends on available energy, this saturation vapor pressure increases with temperature. Table 5-1 and Figure 5-2 show the nonlinear relationship between air temperature and saturation vapor pressure. The *Clausius-Clapeyron* curve in Figure 5-2 depicts the equilibrium point between liquid water and atmospheric water vapor. At this point, the number of molecules evaporating from a flat water surface equals the number condensing on the surface. An air sample below the curve is unsaturated, indicating that evaporation from a flat water surface would exceed condensation onto the surface. The curve shows how higher temperatures (indicating a greater energy supply) increase the potential for evaporation.

Table 5-1
Saturation Vapor Pressure (mb) at Sea-Level Pressure as a Function of Dry-Bulb Temperature (°C)

(°C)	(mb)
-40	0.189
-35	0.314
-30	0.519
-25	0.807
-20	1.254
-18	1.488
-16	1.760
-14	2.076
-12	2.441
-10	2.863
-8	3.348
-6	3.906
-4	4.545
-2	5.275
0	6.108
2	7.055
4	8.129
6	9.347
8	10.722
10	12.272
12	14.017
14	15.977
16	18.173
18	20.630
20	23.373
22	26.430
24	29.831
26	33.608
28	37.796
30	42.430
32	47.551
34	53.200
36	59.422
38	66.264
40	73.777

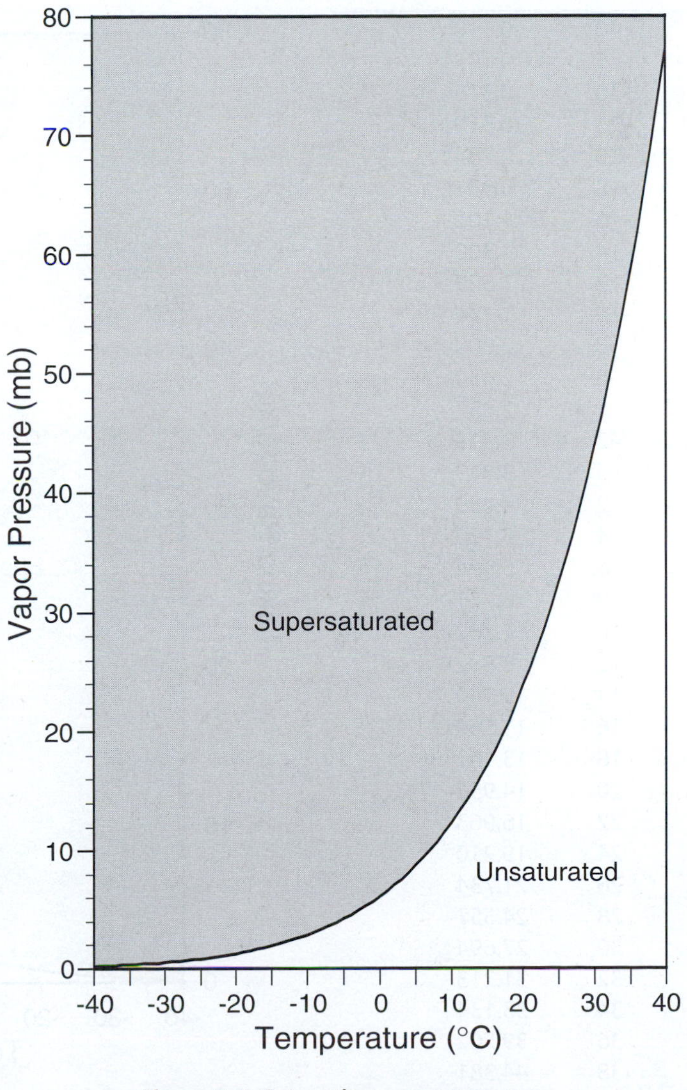

Figure 5-2. Saturation vapor pressure.

Atmospheric water vapor content is measured in other ways. For example, it can also be expressed in terms of the mass of water vapor in the air. The *mixing ratio* is the ratio of water vapor mass to the mass of dry air, usually expressed in units of grams of water vapor per kilogram of dry air (g kg^{-1}). When air is saturated, the ratio is called the *saturation mixing ratio* and represents the maximum weight of water vapor per kilogram of dry air. Like the saturation vapor pressure, this value depends on temperature (Table 5-2). The saturation mixing ratio curve in Figure 5-3 looks similar to the saturation vapor pressure curve, as we have merely substituted one measure of water vapor content for another.

Table 5-2
Saturation Mixing Ratio (g kg^{-1}) at Sea-Level Pressure as a Function of Dry-Bulb Temperature (°C)

(°C)	(g/kg)
-40	0.118
-35	0.195
-30	0.318
-25	0.510
-20	0.784
-18	0.931
-16	1.102
-14	1.300
-12	1.529
-10	1.794
-8	2.009
-6	2.450
-4	2.852
-2	3.313
0	3.819
2	4.439
4	5.120
6	5.894
8	6.771
10	7.762
12	8.882
14	10.140
16	11.560
18	13.162
20	14.956
22	16.963
24	19.210
26	21.734
28	24.557
30	27.694
32	31.213
34	35.134
36	39.502
38	44.381
40	49.815

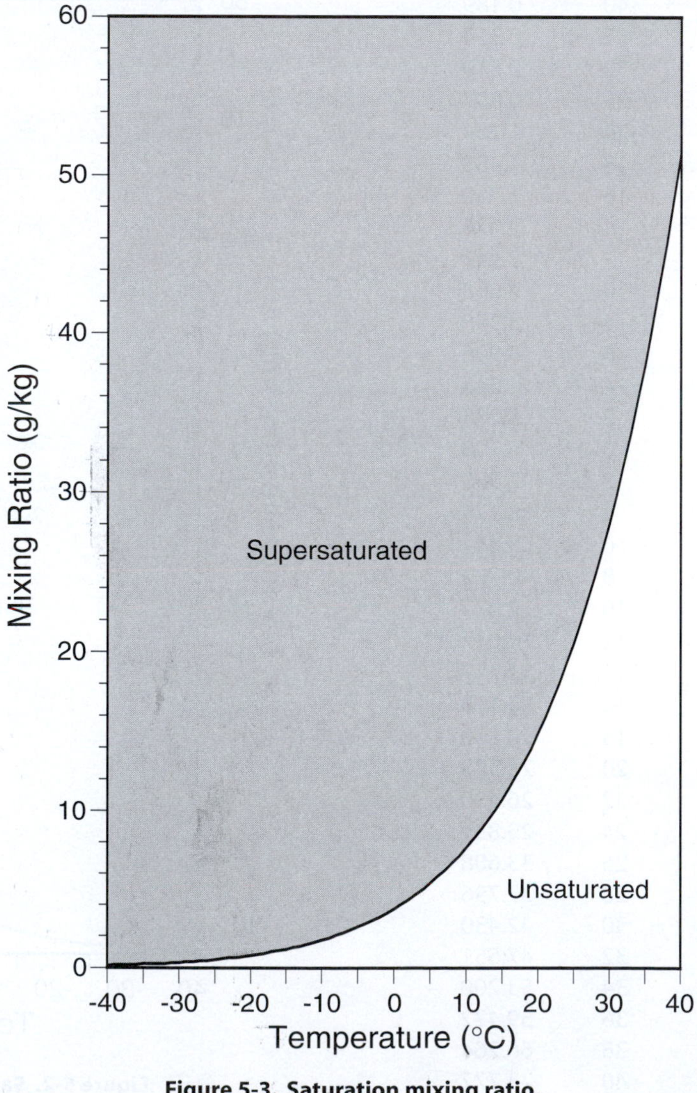

Figure 5-3. Saturation mixing ratio.

Relative Humidity

Vapor pressure, mixing ratio, and the concept of saturation help us to understand a much more commonly used measure of moisture—relative humidity. Remember that a point on the saturation vapor pressure curve represents the vapor pressure that saturated air would exert, and a point on the saturation mixing ratio curve shows the mass of water vapor for a saturated sample. Since the atmosphere typically is not saturated, *relative humidity* (RH) measures how close an air sample is to the saturation point. Specifically, relative humidity is the ratio of the *actual* amount of water vapor in the air to the saturation level at a given temperature. In terms of vapor pressure:

$$RH = \frac{\text{vapor pressure}}{\text{saturation vapor pressure}} \cdot 100\%$$

In terms of mixing ratio:

$$RH = \frac{\text{mixing ratio}}{\text{saturation mixing ratio}} \cdot 100\%$$

Consider a sample of air with a temperature of 20°C and a mixing ratio of 10 g kg^{-1}. The saturation mixing ratio of the sample (derived from Table 5-2) is 14.956 g kg^{-1}. Thus we have:

$$RH = \left[\frac{\text{mixing ratio}}{\text{saturation mixing ratio}}\right] \cdot 100\%$$

$$= \left[\frac{10 \text{ g kg}^{-1}}{14.956 \text{ g kg}^{-1}} \cdot 100\%\right] = 66.8\%$$

14. *Determine the saturation mixing ratio of the following air samples, and calculate the relative humidity of each.*

Temp.	*Saturation Mixing Ratio (g kg^{-1})*	*Mixing Ratio (g kg^{-1})*	*Relative Humidity*
14°C	______	5	______%
14°C	______	9	______%
24°C	______	5	______%
24°C	______	2	______%
34°C	______	7	______%

15. *In the winter, cold air is brought into homes and heated. How does this change the relative humidity of the air?*

16. *Explain why the basement of a house often has high relative humidity in the summer.*

17. *Rank each of the following air samples from 1 (highest) to 5 (lowest) in order of water vapor content.*

	Temp.	*Saturation Mixing Ratio*	*RH*	*Actual Mixing Ratio*	*Rank*
A	14°C	______	90%	______	______
B	20°C	______	60%	______	______
C	24°C	______	40%	______	______
D	30°C	______	40%	______	______
E	34°C	______	30%	______	______

18. *From the above example, can you suggest one disadvantage to using relative humidity?*

Table 5-3
Relative Humidity Table (°F)

Dry-Bulb Temp (°F)	Wet-Bulb Depression (°F) 1	2	3	4	5	6	7	8	9	10	11	12	13	14	15	16	17	18	19	20	21	22	23	24	25	30	35
0	67	33	1																								
5	73	46	20																								
10	78	56	34	13																							
15	82	64	46	29	11																						
20	85	70	55	40	26	12																					
25	**87**	**74**	**62**	**49**	**37**	**25**	**13**	**1**																			
30	89	78	67	56	46	36	26	16	6																		
35	91	81	72	63	54	45	36	27	19	10	2																
40	92	83	75	68	60	52	45	37	29	22	15	7															
45	93	86	78	71	64	57	51	44	38	31	25	18	12	6													
50	**93**	**87**	**80**	**74**	**67**	**61**	**55**	**49**	**43**	**38**	**32**	**27**	**21**	**16**	**10**	**5**											
55	94	88	82	76	70	65	59	54	49	43	38	33	28	23	19	11	9	5									
60	94	89	83	78	73	68	63	58	53	48	43	39	34	30	26	21	17	13	9	5	1						
65	95	90	85	80	75	70	66	61	56	52	48	44	39	35	31	27	24	20	16	12	9	5	2				
70	95	90	86	81	77	72	68	64	59	55	51	48	44	40	36	33	29	25	22	19	15	12	9	6	3		
75	**96**	**91**	**86**	**82**	**78**	**74**	**70**	**66**	**62**	**58**	**54**	**51**	**47**	**44**	**40**	**37**	**34**	**30**	**27**	**24**	**21**	**18**	**15**	**12**	**9**		
80	96	91	87	83	79	75	72	68	64	61	57	54	50	47	44	41	38	35	32	29	26	23	20	18	15	3	
85	96	92	88	84	81	77	73	70	66	63	59	57	53	50	47	44	41	38	36	33	30	27	25	22	20	8	
90	96	92	89	85	81	78	74	71	68	65	61	58	55	52	49	47	44	41	39	36	34	31	29	26	24	13	3
95	96	93	89	86	82	79	76	73	69	66	63	61	58	55	52	50	47	44	42	39	37	34	32	30	28	17	8
100	**96**	**93**	**89**	**86**	**83**	**80**	**77**	**73**	**70**	**68**	**65**	**62**	**59**	**56**	**54**	**51**	**49**	**46**	**44**	**41**	**39**	**37**	**35**	**33**	**30**	**21**	**12**
105	97	93	90	87	84	81	78	75	72	69	66	64	61	58	56	53	51	49	46	44	42	40	38	36	34	24	15
110	97	93	90	87	84	81	78	75	73	70	67	65	62	60	57	55	52	50	48	46	44	42	40	38	36	26	18
115	97	94	91	88	85	82	79	76	74	71	69	66	64	61	59	57	54	52	50	48	46	44	42	40	38	29	21
120	97	94	91	88	85	82	80	77	74	72	69	67	65	62	60	58	55	53	51	49	47	45	43	41	40	31	23
125	**97**	**94**	**91**	**88**	**86**	**83**	**80**	**78**	**75**	**73**	**70**	**68**	**66**	**64**	**61**	**59**	**57**	**55**	**53**	**51**	**49**	**47**	**45**	**44**	**42**	**33**	**26**
130	97	94	91	89	86	83	81	78	76	73	71	69	67	64	62	60	58	56	54	52	50	48	47	45	43	35	28

Table 5-4
Relative Humidity Table (°C)

Dry-Bulb Temp (°C)	**Wet-Bulb Depression (°C)** 1	2	3	4	5	6	7	8	9	10	11	12	13	14	15	16	17	18	19	20	21	22
-20	**28**																					
-10	**66**	**33**	**0**																			
-8	71	41	13																			
-6	73	48	20	0																		
-4	77	54	32	11																		
-2	79	58	37	20	1																	
0	**81**	**63**	**45**	**28**	**11**																	
2	83	67	51	36	20	6																
4	85	70	56	42	27	14																
6	86	72	59	46	35	22	10	0														
8	87	74	62	51	39	28	17	6														
10	**88**	**76**	**65**	**54**	**43**	**33**	**24**	**13**	**4**													
12	88	78	67	57	48	38	28	19	10	2												
14	89	79	69	60	50	41	33	25	16	8	1											
16	90	80	71	62	54	45	37	29	21	14	7	1										
18	91	81	72	64	56	48	40	33	26	19	12	6	0									
20	**91**	**82**	**74**	**66**	**58**	**51**	**44**	**36**	**30**	**23**	**17**	**14**	**5**									
22	92	83	75	68	60	53	46	40	33	27	21	15	10	4	0							
24	92	84	76	69	62	55	49	42	36	30	25	20	14	9	4	0						
26	92	85	77	70	64	57	51	45	39	34	28	23	18	13	9	5						
28	93	86	78	71	65	59	53	45	42	36	31	26	21	17	12	8	4					
30	**93**	**86**	**79**	**72**	**66**	**61**	**55**	**49**	**44**	**39**	**34**	**29**	**25**	**20**	**16**	**12**	**8**	**4**				
32	93	86	80	73	68	62	56	51	46	41	36	32	27	22	19	14	11	8	4			
34	93	86	81	74	69	63	58	52	48	43	38	34	30	26	22	18	14	11	8	5		
36	94	87	81	75	69	64	59	54	50	44	40	36	32	28	24	21	17	13	10	7	4	
38	94	87	82	76	70	66	60	55	51	46	42	38	34	30	26	23	20	16	13	10	7	5
40	**94**	**89**	**82**	**76**	**71**	**67**	**61**	**57**	**52**	**48**	**44**	**40**	**36**	**33**	**29**	**25**	**22**	**19**	**16**	**13**	**10**	**7**

Despite its limitations, we can easily measure relative humidity. In fact, accurate measurement of RH allows you to compute other measures such as vapor pressure and mixing ratio (which are *not* easy to measure). The *sling psychrometer* provides one means of measuring relative humidity. This simple instrument consists of two thermometers mounted on a swinging device. The bulb of one thermometer is covered with gauze that is saturated with water. The other thermometer is dry. As you twirl the mounted thermometers, water evaporates from the wet bulb, and energy is taken away from it. The principle of the sling psychrometer is the same as illustrated in Experiment II and the phase change question earlier—evaporation consumes energy. The *wet-bulb temperature* can never be greater than the *dry-bulb temperature*. If the air is very moist, little liquid water will evaporate from the wet bulb and the difference between the wet-bulb and dry-bulb temperatures (the *wet-bulb depression*) will be small. If the air is dry, more water evaporates from the wet bulb, and the wet-bulb depression will be greater.

19. *Work through the following two examples using Table 5-4 (on the previous page).*

	A	B
Dry-bulb temperature	32°C	10°C
Wet-bulb temperature	25°C	5°C
Wet-bulb depression	________	________
Relative humidity	________	________

Optional Exercise: Measuring Relative Humidity

20. *Use a sling psychrometer and Table 5-3 or 5-4 to determine the relative humidity of the air indoors and at three sites around campus that you think will have different relative humidities.*

	Dry-Bulb Temp.	Saturation Mixing Ratio	Wet-Bulb Temp.	Wet-Bulb Depression	Relative Humidity	Mixing Ratio
Indoors	______°	______ g kg^{-1}	______°	______°	______%	______ g kg^{-1}
Site 1 Description:	______°	______ g kg^{-1}	______°	______°	______%	______ g kg^{-1}
Site 2 Description:	______°	______ g kg^{-1}	______°	______°	______%	______ g kg^{-1}
Site 3 Description:	______°	______ g kg^{-1}	______°	______°	______%	______ g kg^{-1}

21. *Was there a difference between the site with the highest relative humidity and the site with the highest mixing ratio? If so, explain what might have contributed to this difference. Explain any other observed differences among the three sites. Do your results make sense to you? What sources of error could have influenced your measurements?*

Dew Point

Dew point is another important measure of atmospheric moisture. Consider the sample of air in Figure 5-4 with a temperature of 20°C and mixing ratio of 10 g kg^{-1}. Its *dew point* is the temperature to which it must be cooled in order to reach saturation. We may illustrate this idea on the graph by simply plotting the value of temperature and mixing ratio of the unsaturated air parcel and drawing a horizontal line that intersects the saturation curve. The temperature at the point of intersection is the dew-point temperature (14°C here). We commonly witness this process when air near the earth's surface is chilled to its saturation point and dew forms on the ground.

22. *What is the dew point of an air sample with a temperature of 30°C and a mixing ratio of 17 g kg^{-1}?*

Remember that air is saturated at the dew point. Use Table 5-2 or Figure 5-3 to confirm that the mixing ratio from question 22 (17 g kg^{-1}) is the *saturation* mixing ratio for the dew point that you derived from Figure 5-4. This illustrates how temperature relates to saturation mixing ratio and how dew point relates to mixing ratio.

23. *Now consider an air sample with a temperature of 34°C and a relative humidity of 62%.*
    What is its saturation mixing ratio? ________
    What is its mixing ratio? ________
    Plot the air sample on Figure 5-4.
    What is its dew point? ________

Optional Exercise

24. *Determine the dew point of one or two sites you measured in question 20.*

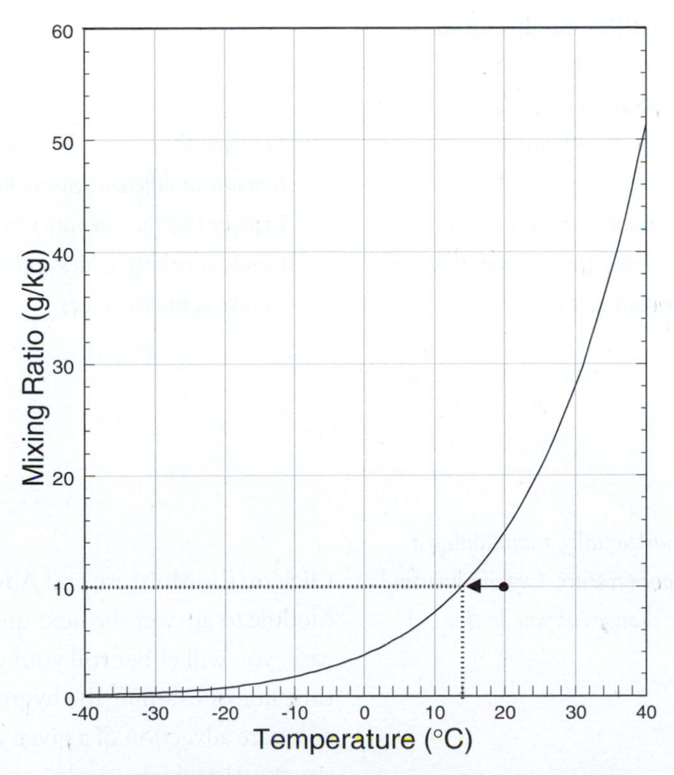

Figure 5-4. Dew point of an unsaturated air sample.

Optional Exercise: Measuring Dew Point

25. *Fill a container with water. Measure the air temperature in the room (__________ °C), then place the thermometer in the container of water. Slowly add ice and stir. When water (dew) forms on the container the dew-point temperature has been reached. Record this dew-point temperature: __________ °C. Using the air temperature, determine the saturation mixing ratio in the room: __________ g kg^{-1}. Calculate the relative humidity in the room and compare this value with the relative humidity measured with a sling psychrometer. Note that this experiment will not work if the dew point is less than 0°C (32°F). Can you explain why?*

Interactive Computer Exercise: Moisture

This module focuses on measures of moisture. The first part allows you to adjust temperature and moisture content of an air sample to review the concept of saturation. The second part shows how advection of different air masses affects temperature and humidity. Start the module and click on the Saturation Vapor Pressure tab to answer the questions below.

26. *What variable must you alter on the graph to change the evaporation rate? Explain why this variable influences evaporation rate.*

27. *Slide the air sample horizontally, maintaining a relatively constant vapor pressure. Explain how and why relative humidity changes as you do this.*

28. *Now slide the air sample vertically, maintaining a relatively constant temperature. Explain how and why relative humidity changes as you do this.*

29. *Compare the evaporation rate with the condensation rate at different relative humidity values. Explain why the rates are very different at low relative humidity values and very similar at high relative humidity values.*

Click on the Moisture and Advection tab of the module to answer the next questions. In each case, you will either roll your cursor over the time line or examine the hygrothermographs to compare advection of a given air mass against a situation in which no advection occurs.

30. *Advection of which air mass(es) produces higher afternoon temperatures in Kansas City?*

31. a. *Which air mass(es) causes higher afternoon relative humidity in Kansas City?*

    b. *Which causes higher dew point in Kansas City?*

    c. *How could the advection of a given air mass lead to both higher relative humidity and lower dew point?*

Review Questions

Compare and contrast two different measures of humidity. What is the basis of each measure? What are the strengths and limitations of each?

Why does your soft-drink can "sweat" more in the summer and than in the winter?

How would you describe the humidity of air circulated in the cabin of a jet flying at a height of 11 km if this air is drawn from outside the aircraft and warmed to 22°C? What effect(s) might this have on your body?

How is the principle of the sling psychrometer similar to the process by which the human body cools itself through sweating?

Lab 6

SATURATION AND ATMOSPHERIC STABILITY

Materials Needed

- calculator
- ruler

Introduction

How do clouds form? The concept of saturation, introduced in the previous lab, is important here. The two basic means by which air achieves saturation (relative humidity = 100%) are cooling and increasing water vapor content. In this lab we will examine several common atmospheric processes that will lead to either or both of these phenomena. We will focus particularly on how air changes as it rises from the earth's surface and how these changes relate to atmospheric stability.

Mixing

Saturation occasionally occurs when two air masses mix. Mixing usually involves a change in both water vapor content and temperature. Consider, for example, two unsaturated air samples, A and B:

Air Sample	**A**	**B**
Temperature	10°C	38°C
Relative humidity	75%	75%

1. *Use Table 6-1 to find the saturation mixing ratio (m_s) of air samples A and B.*

   A: m_s = ________ $g\ kg^{-1}$

   B: m_s = ________ $g\ kg^{-1}$

2. *What are the mixing ratio (m) values of samples A and B?*

   A: m = ________ $g\ kg^{-1}$

   B: m = ________ $g\ kg^{-1}$

3. *Use the temperature and mixing ratio data to plot the two air samples in Figure 6-1.*

Assume that equal parts of these two air masses mix together. The resulting temperature is halfway between 10°C and 38°C, and the mixing ratio is halfway between the values that you found for the two samples.

Table 6-1
Saturation Mixing Ratio (g kg^{-1})
at Sea-Level Pressure
as a Function of Dry-Bulb
Temperature (°C)

(°C)	(g/kg)
-40	0.118
-35	0.195
-30	0.318
-25	0.510
-20	0.784
-18	0.931
-16	1.102
-14	1.300
-12	1.529
-10	1.794
-8	2.009
-6	2.450
-4	2.852
-2	3.313
0	3.819
2	4.439
4	5.120
6	5.894
8	6.771
10	7.762
12	8.882
14	10.140
16	11.560
18	13.162
20	14.956
22	16.963
24	19.210
26	21.734
28	24.557
30	27.694
32	31.213
34	35.134
36	39.502
38	44.381
40	49.815

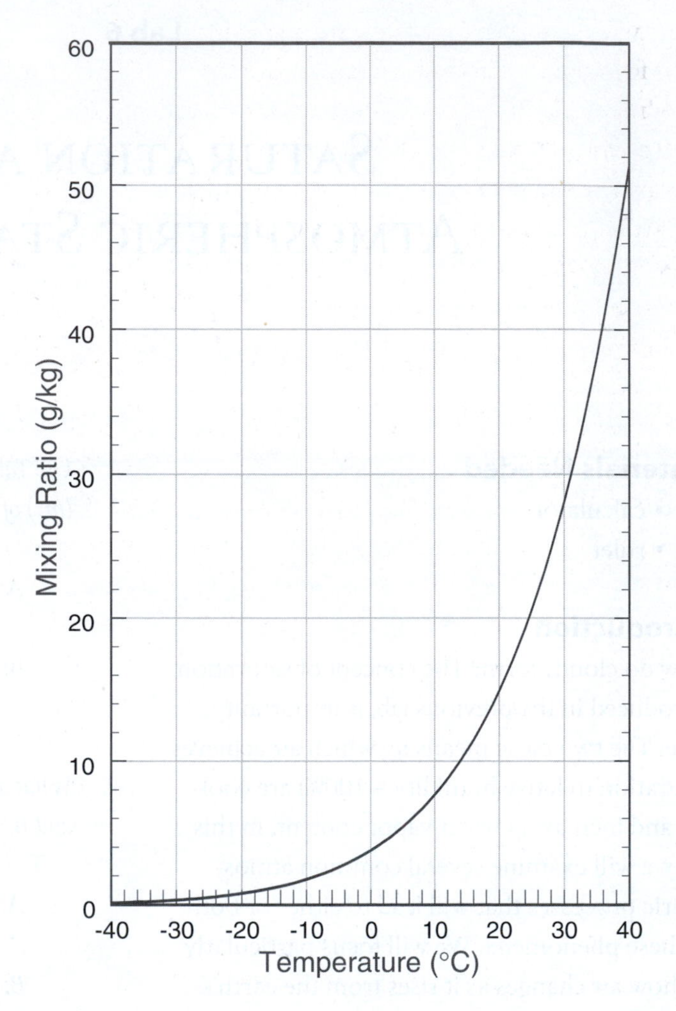

Figure 6-1

4. *Plot the mixed air sample in Figure 6-1.*

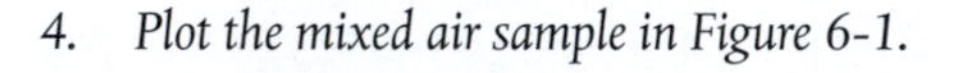

5. *What is the new temperature?* __________

6. *Saturation mixing ratio?* __________

7. *Mixing ratio?* __________

8. *Relative humidity?* __________

9. *The nonlinear relationship between temperature and saturation makes it possible for two unsaturated air samples to mix and form a new saturated air sample. Jet contrails are one example of this as warm exhaust containing moisture mixes with cold air. Some contrails disappear quickly, others linger. What do you think determines the life span of a contrail?*

Adiabatic Cooling

Saturation occurs more commonly when air is cooled to its dew-point temperature than when air masses mix. In nature, this cooling often occurs when air rises. Consider a parcel of air that is forced to rise and does not mix with the surrounding environment. As the parcel rises atmospheric pressure around it decreases, allowing the parcel to expand (Figure 6-2). Since the parcel has the same number of molecules but occupies more volume, its average internal energy (i.e., temperature) decreases.

Air can be forced to rise because of:

a. Intense surface heating.

b. The "collision" or convergence of surface air masses.

c. The contact of air masses of unlike temperature along warm and cold fronts.

d. A topographic barrier (such as a mountain range).

e. Upper-air divergence (a concept we will discuss in a later lab).

As an unsaturated air parcel rises its temperature will decrease at the *dry adiabatic lapse rate* (DALR) of approximately 10°C per kilometer.

10. *Calculate the temperature of an unsaturated air parcel at 100-m increments as it is forced to rise from the earth's surface, where its temperature is 35°C.*

Height (m)	*Temperature*
1000 (1 km)	______
900	______
800	______
700	______
600	______
500	______
400	______
300	______
200	______
100	______
Earth's surface	*35°C*

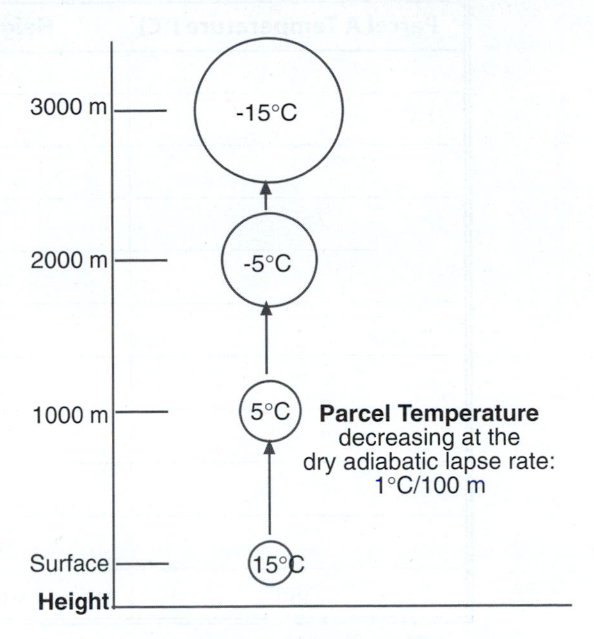

Figure 6-2. Adiabatic cooling of an unsaturated air parcel.

It is possible that rising air parcels will cool to the dew-point temperature. When this happens, a parcel becomes saturated and condensation or clouds form. The height at which this occurs is called the *lifting condensation level* (LCL). If the air parcel rises above the LCL, it cools at a slower rate referred to as the *saturated, moist* or *wet adiabatic lapse rate* (WALR). The wet adiabatic lapse rate ranges between 5°C and 9°C per kilometer. It is slower than the dry adiabatic lapse rate because latent heat is released within the parcel as water vapor condenses, a warming that partially offsets adiabatic cooling. The WALR varies because the amount of condensation depends both on the amount of water vapor in the parcel and on atmospheric pressure.

Table 6-2

Parcel A Temperature (°C)	Height (km)	Parcel B Temperature (°C)
	5.0	
	4.5	
	4.0	
	3.5	
	3.0	
	2.5	
	2.0	
	1.5	
	1.0	
	0.5	
28°C	surface	10°C

11. *Consider air parcel A, which is 28°C at the surface. It is forced to rise to 5 km. The lifting condensation level is 1.5 km, above which the parcel cools at an average wet adiabatic lapse rate of 5°C per kilometer. Fill in the left column of Table 6-2 indicating the parcel's changing temperature.*

12. *Now consider air parcel B with a surface temperature of 10°C that is forced to rise. It too reaches the LCL at 1.5 km, but the average wet adiabatic lapse rate in this case is 7°C per kilometer. Fill in the right column of the table indicating temperature change in the lower 5 kilometers.*

13. *Why would the warmer parcel cool at a slower rate between 1.5 and 5 km?*

In the previous example, the lifting condensation level was provided for you. We can calculate the LCL from surface temperature and dew-point temperature. The LCL generally occurs at the height where air has cooled to the dew-point temperature. Like temperature, dew-point temperature also decreases with height because of decreasing pressure. As a rule of thumb, dew point drops approximately 2°C per kilometer.

Consider the example shown in Figure 6-3. An air parcel at sea level has the following characteristics: **Temperature = 30°C, Dew-point temperature = 14°C, Pressure = 1010 mb.** If the parcel is forced to rise, its temperature drops at the dry adiabatic lapse rate (10°C per kilometer) until it becomes saturated. The drop in dew-point temperature (2°C per kilometer) can be found by following a *mixing ratio line* associated with dew-point temperature at the surface. The point where the lines intersect is the lifting condensation level.

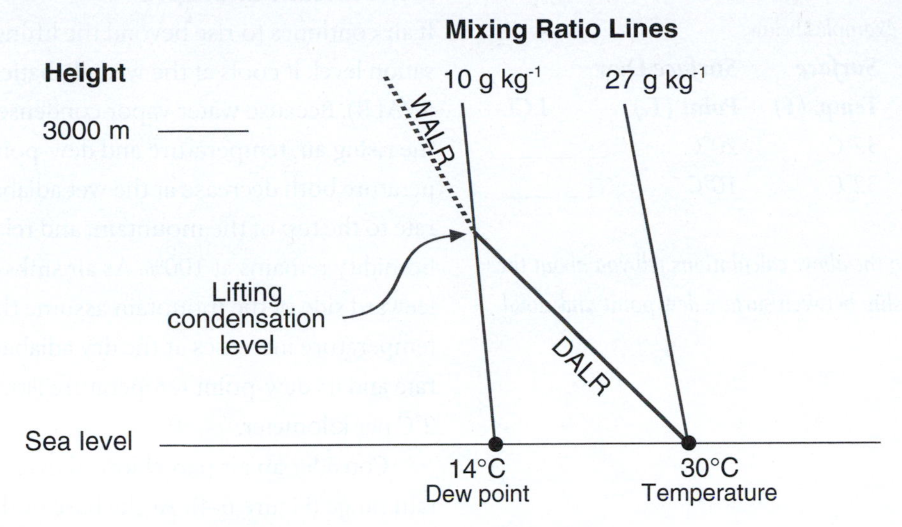

Figure 6-3. Lifting condensation level.

14. *Fill in Table 6-3 below to show the temperature and dew-point temperature of the forced air parcel.*

If we know the surface temperature and dew point of an air parcel, we can use the following equation to calculate the point of intersection:

$$T_d - 2x = T - 10x$$

The left side of the equation represents the decrease of dew-point temperature with height; the right side represents the temperature decrease with height. At some height x these expressions are equal. Solving algebraically, we find:

$$10x - 2x = T - T_d$$

$$8x = 30 - 14$$

$$8x = 16$$

Now, recalling the units in our equation:

$$x = \frac{16°C}{8°C\ km^{-1}}$$

$$x = 2\ km$$

Now that you understand the concept, here's a shortcut:

$$LCL\ (km) = \frac{T(°C) - T_d(°C)}{8°C\ km^{-1}}$$

or

$$LCL\ (meters) = 125\ [T(°C) - T_d(°C)]$$

Table 6-3

Height (m)	Temperature (T°C)	Dew-Point Temperature (T_d°C)
2000		
1500		
1000		
500		
earth's surface	30°C	14°C

15. *Calculate the lifting condensation level (LCL) for the two examples below.*

	Surface Temp. (T)	*Surface Dew Point (T_d)*	LCL
Example A	*32°C*	*20°C*	______
Example B	*32°C*	*10°C*	______

16. *What do the above calculations tell you about the relationship between surface dew point and cloud height?*

Optional Exercise

17. *On a day with convective cumulus clouds, measure the outdoor temperature and dew-point temperature and determine the approximate height of the cloud bases.*

A Mountain Example

If air continues to rise beyond the lifting condensation level, it cools at the wet adiabatic lapse rate (WALR). Because water vapor condenses out of the rising air, temperature and dew-point temperature both decrease at the wet adiabatic lapse rate to the top of the mountain, and relative humidity remains at 100%. As air sinks on the leeward side of the mountain assume that its temperature increases at the dry adiabatic lapse rate and its dew-point temperature *increases* at 2°C per kilometer.

Consider an air parcel forced over a mountain range (Figure 6-4). At the base of the windward side of the mountain the temperature is 25°C and the dew-point temperature is 13°C.

18. *Fill in the spaces for temperature and dew-point temperature at various heights on the windward and leeward sides of the mountain. Assume that the wet adiabatic lapse rate is 5°C per kilometer. (Note that temperature and dew point drop together when a saturated air parcel rises and water vapor condenses.)*

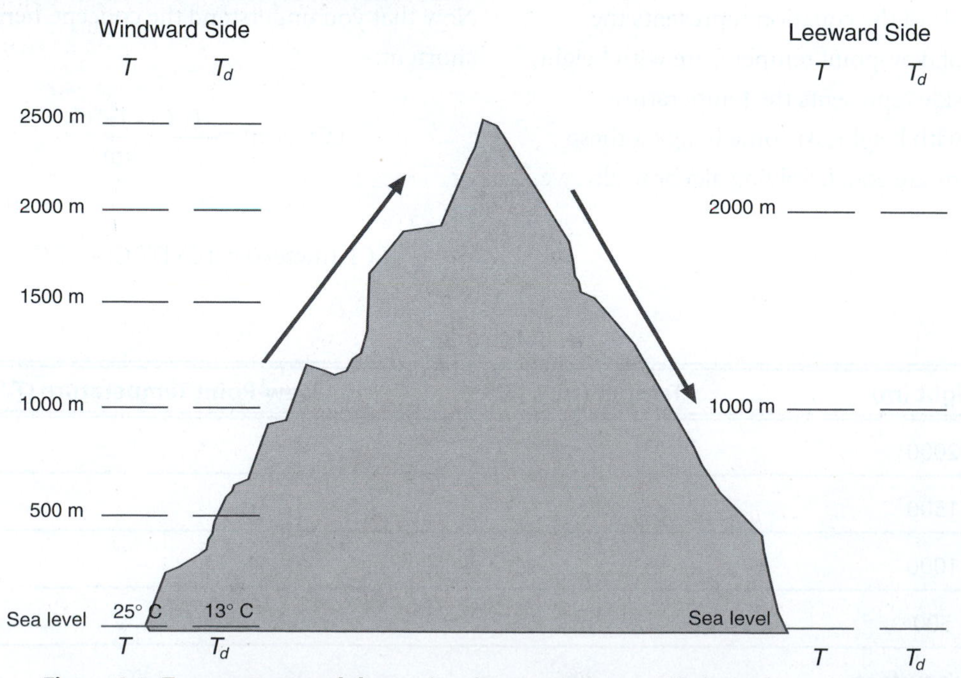

Figure 6-4. Temperature and dew point change as air parcel is forced over a mountain.

19. *At what height would cloud bases form?*

__________ *meters*

20. *How do the sea-level temperature and dew point on the leeward side compare with the sea-level temperature and dew point on the windward side?*

21. *Which side of the mountain is more often cloudy and which side is more often clear?*

Atmospheric Stability

In our discussion so far we have made two assumptions: (a) that certain processes force air to rise from the earth's surface and (b) that rising air does not mix substantially with the surrounding atmosphere. Once an initial lifting force ceases, the fate of a rising air parcel depends on the state of the atmosphere through which it rises, or *atmospheric stability.* Consider the three examples in Figures 6-5 through 6-7. In each example the environmental temperature is shown with a solid line, and the parcel temperature follows the dry and wet adiabatic rates of cooling (dashed lines). In each example, the surface temperature is 30°C, surface dew point is 14°C, and the wet adiabatic lapse rate is 5°C/km.

Weather balloons measure the temperature above the earth's surface. These measurements allow us to calculate the rate at which the surrounding environmental temperature changes with height—the *environmental lapse rate* (ELR). The atmosphere is considered *absolutely stable* if an air parcel that is forced aloft cools faster than the surrounding environment (Figure 6-5), that is, when the parcel rate of cooling (the dry or wet adiabatic lapse rate) is faster than the environmental lapse rate. If the force lifting the parcel continued to act, condensation would eventually occur, but clouds would be layered clouds without much vertical development. If the force ceased, the parcel would have a tendency to sink.

In an *absolutely unstable* atmosphere a parcel that is initially displaced will continue to rise because it is warmer and more buoyant than its surroundings at the displaced height. In this case, the parcel rate of cooling is slower than the

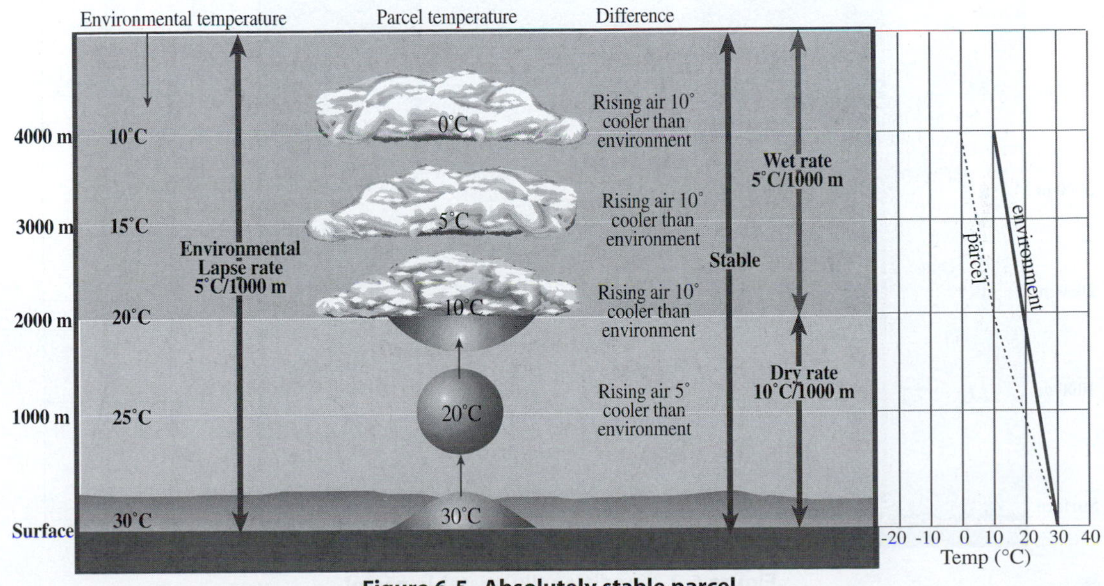

Figure 6-5. Absolutely stable parcel.

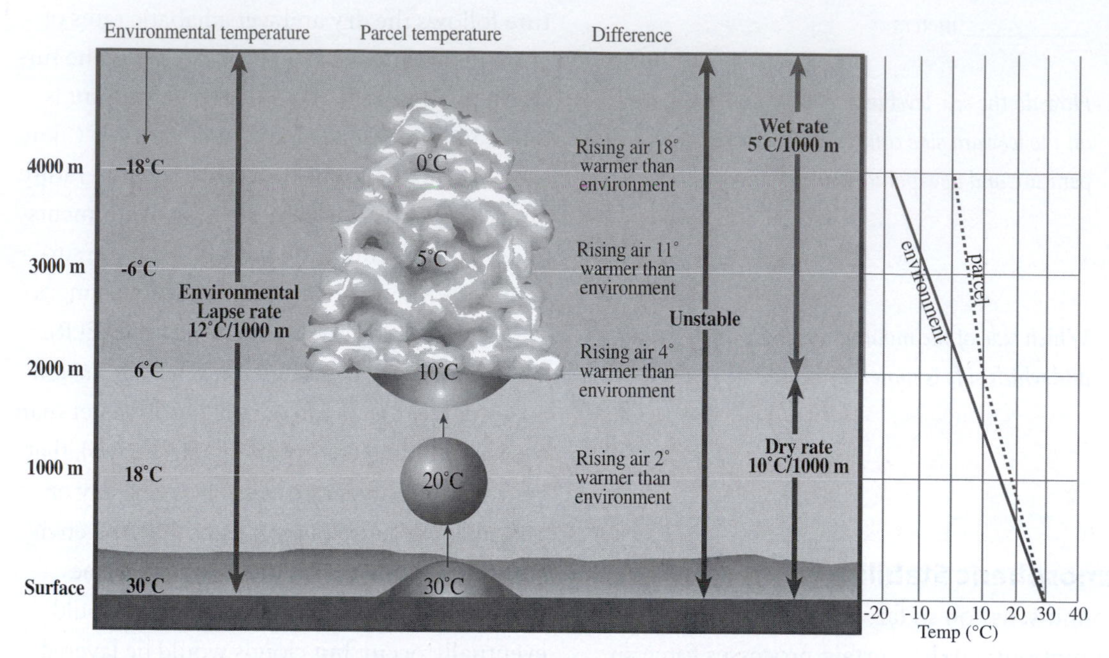

Figure 6-6. Absolutely unstable parcel.

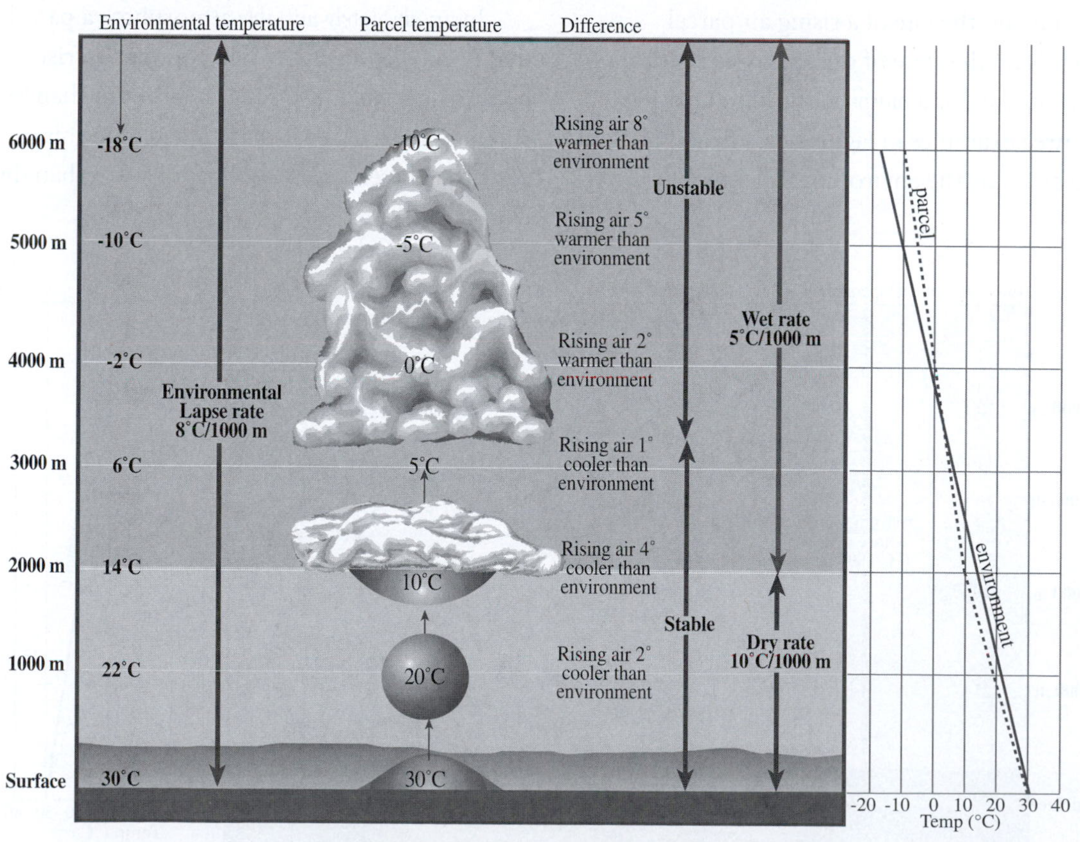

Figure 6-7. Conditionally unstable parcel.

environmental lapse rate. If the air parcel rises to its lifting condensation level, clouds with vertical development will form as the buoyant air rises on its own accord (Figure 6-6). Thunderstorms require such conditions.

A more complex situation is depicted in Figure 6-7, where the lower portion of the atmosphere is stable (the parcel temperature is cooler than the surrounding environment), and higher elevations are unstable (the parcel temperature is warmer than the surrounding environment). In this situation the atmosphere can be described as *conditionally unstable.* It exists, of course, because the wet and dry adiabatic lapse rates are different, and the environmental lapse rate is between the two (DALR>ELR>WALR). Again, notice that the resulting cloud patterns reflect the tendency for vertical development only in unstable layers of the atmosphere (above 3300 meters in this example).

Collectively, the three examples illustrate how different stability conditions affect cloud formation. In each case, the surrounding environmental temperature determines atmospheric stability and the fate of a rising air parcel.

Using one final example, let us examine how a meteorologist would measure atmospheric instability and determine whether to issue a thunderstorm forecast. Consider an air parcel that rises from the surface to 3 kilometers. The surface temperature is 31°C, dew-point temperature 23°C, and pressure 1010 mb. Table 6-4 lists the measured temperature of the surrounding atmosphere through which the parcel rises.

22. *What is the lifting condensation level in this example? __________ km*

23. *Enter the correct parcel temperatures and parcel dew points to complete Table 6-4. (Assume that the WALR = 5°C/km).*

Table 6-4

Height	Environmental Temp.	Parcel Temp.	Parcel Dew Point
8.0 km	-28°C	-14°C	-14°C
7.5 km	-25°C	-11.5°C	-11.5°C
7.0 km	-22°C	-9°C	-9°C
6.5 km	-19°C	-6.5°C	-6.5°C
6.0 km	-16°C	-4°C	-4°C
5.5 km	-13°C	-1.5°C	-1.5°C
5.0 km	-10°C	1°C	1°C
4.5 km	-7°C	3.5°C	3.5°C
4.0 km	-3°C	6°C	6°C
3.5 km	1.5°C		
3.0 km	6°C		
2.5 km	10.5°C		
2.0 km	15°C		
1.5 km	19°C		
1.0 km	23°C		
0.5 km	27°C		
surface	31°C	31°C	23°C

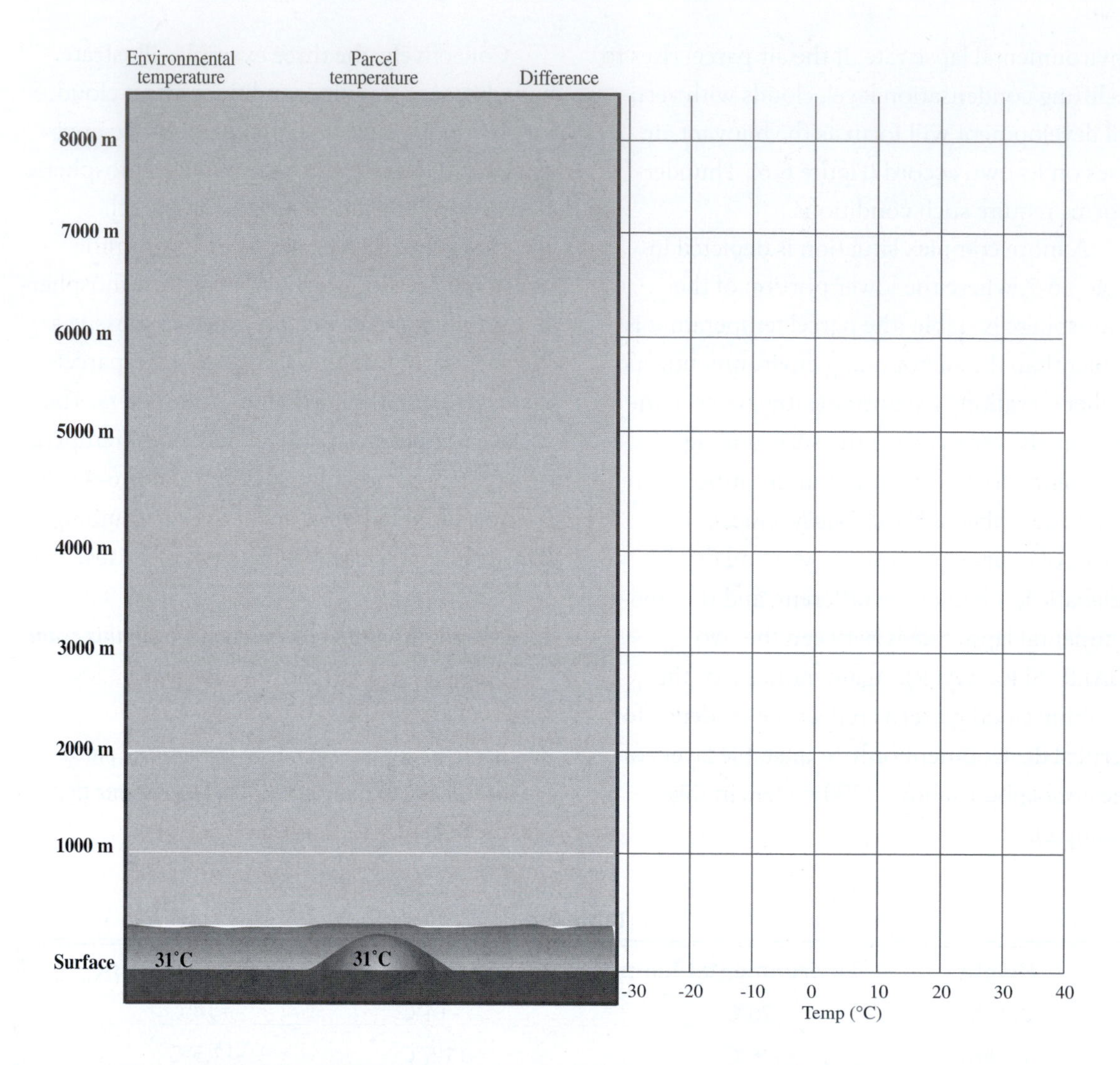

24. *Using the diagram above and Figures 6-5 through 6-7 as models:*
    a. *Plot the parcel and environmental temperatures from Table 6-3.*
    b. *Show the lifting condensation level and draw likely cloud development.*
    c. *Identify any stable or unstable layers.*

25. *Would you forecast thunderstorms under these conditions?*

Review Questions

How is saturation achieved to form most clouds?

What is the difference between a stable and an unstable atmosphere? Which of these two conditions favors vertical cloud development and why?

Lab 7

Cloud Droplets and Raindrops

Materials Needed

- calculator

Introduction

How do raindrops and snowflakes form? This lab examines the forces acting on cloud droplets and ice crystals and the processes that cause these droplets and crystals to grow. We will consider how such growth influences the probability that clouds will produce precipitation.

Cloud Droplet Growth

Clouds form as water vapor either condenses on condensation nuclei or is deposited on freezing nuclei. Typically, cloud droplets and ice crystals are very small (50 μm = 0.05 mm). For clouds to produce precipitation, cloud droplets must grow large enough and fast enough to fall to the ground without evaporating. Droplets reach this size through two processes: the *Bergeron process* and the *collision-coalescence process.* The Bergeron process of ice-crystal growth depends on the coexistence of water vapor, supercooled liquid water droplets, and ice. To understand the importance of this coexistence, consider again the Clausius-Clapeyron (saturation vapor pressure) curve in Figure 7-1. Remember that the solid curve depicts the equilibrium point between liquid water and atmospheric water vapor over a range of temperatures. An air sample below the curve is considered unsaturated, indicating that evaporation exceeds condensation, and an air sample above the curve is considered *supersaturated,* suggesting that condensation exceeds evaporation.

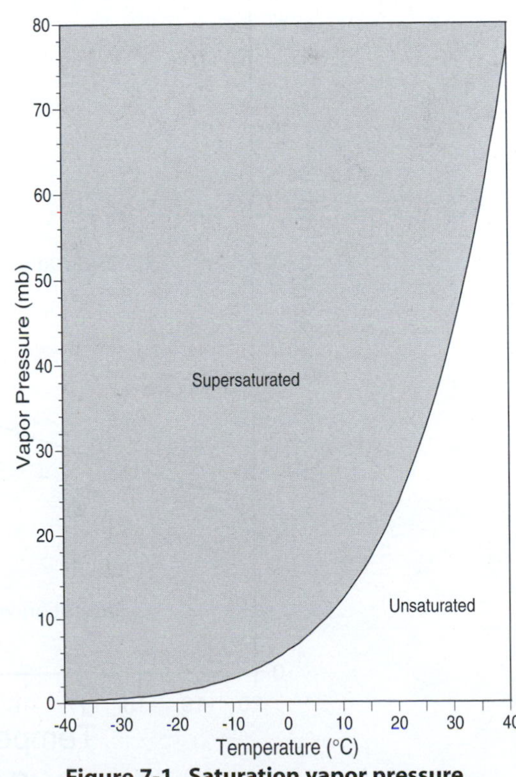

Figure 7-1. Saturation vapor pressure.

Clouds with temperatures between -10° and -20°C often contain water in all three forms—water vapor, supercooled liquid water droplets, and ice crystals. The tendency for water to move from one phase to another depends on the equilibrium between phases. Since water molecules are more tightly bonded in solid than in liquid form, more energy is required to change the phase of water from solid to vapor (a process called *sublimation*) than from liquid to vapor (evaporation). Therefore, the saturation vapor pressure over liquid water is greater than the saturation vapor pressure over ice, and it is possible for an air sample to be unsaturated with respect to liquid water and supersaturated with respect to ice (Figure 7-2).

1. *Consider the air sample denoted by an asterisk (*) in Figure 7-2. Its temperature is -15°C and its vapor pressure is 1.72 mb. What is its relative humidity with respect to water if, at -15°C, the saturation vapor pressure over liquid water is 1.91 mb? (Your answer should be less than 100%, indicating an unsaturated sample.)*

2. *What will likely happen to liquid water when the surrounding environment is unsaturated?*

3. *What is its relative humidity with respect to ice if, at -15°C, the saturation vapor pressure over ice is only 1.65 mb?*

4. *What happens to water vapor if, with respect to the surrounding ice crystals, the environment is supersaturated?*

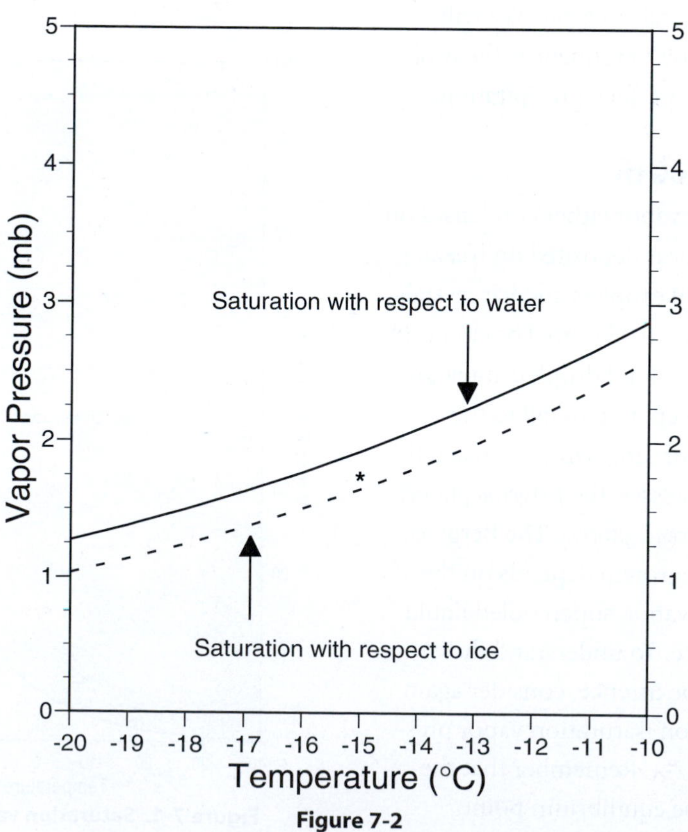

Figure 7-2

Forces Acting on Cloud Droplets and Raindrops

Two opposing forces act on droplets—*gravity* and *friction*. Gravitational force pulls a water droplet to the earth's surface and is equal to the mass of the droplet times gravitational acceleration (gravity = mass $\bullet 9.8 \text{ m} \bullet \text{s}^{-2}$). Since $1 \text{ cm}^3 = 1 \text{ g}$, we can substitute volume for mass in the above equation. Therefore, the gravitational force of a droplet increases with its volume. The volume of a sphere is calculated as follows:

$$V = \frac{4}{3}\pi r^3$$

where *r* is the radius of the droplet.

5. *What is the volume of an ordinary cloud droplet (25 µm = 0.025 mm radius)?*

6. *A large raindrop (2500 µm = 2.5 mm radius) is 100 times the radius of a typical cloud droplet (25 µm). How many times greater is its volume?*

As a droplet falls it encounters air resistance, or frictional force. The magnitude of this force depends on its speed and the size of the drop's "bottom"—i.e., the surface area resisting the fall. Figure 7-3 illustrates the "bottom" of a cloud droplet with a 25-µm spherical radius. Assuming the drop is spherical, the surface area experiencing friction is that of a circle, calculated as: Area = πr^2.

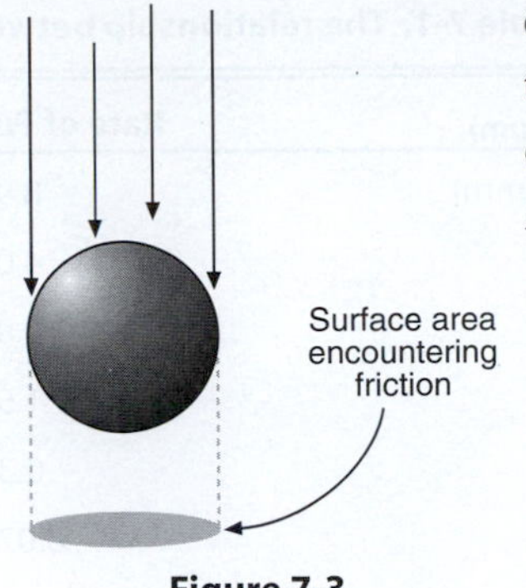

Figure 7-3

7. *What is the area of the "bottom" of a falling small raindrop with a 0.5-mm radius?*

8. *How many times greater is the "bottom" area of a falling large droplet with a 2.5-mm radius?*

9. *Since gravitational force is proportional to the mass (or volume) of a raindrop, and frictional force is proportional to the area of the droplet encountering resistance, which of the two forces increases more for a given increase in droplet radius?*

Terminal Velocity

As gravity first acts on a droplet, the droplet accelerates. In turn, frictional drag increases as a droplet accelerates, since a faster droplet encounters more air molecules. Eventually frictional and gravitational forces balance and the droplet no longer accelerates, but falls at constant speed. This speed is referred to as the droplet's *terminal velocity*. Your calculations above show that as droplet size increases, gravitational force is affected more than frictional force.

Table 7-1. The relationship between droplet size and terminal velocity.

Radius (μm)	Rate of Fall (m s-¹)	Type of Drop
2500 (2.5 mm)	8.9	Large raindrop
500	4.0	Small raindrop
250	2.8	Fine rain or large drizzle
100	1.5	Drizzle
50	0.3	Large cloud droplet
25	0.076	Ordinary cloud droplet
5	0.003	Small cloud droplet
0.5	0.00004	Large condensation nucleus

Therefore, smaller droplets will have a lower terminal velocity than larger droplets (Table 7-1). Terminal velocity is important because the speed of a droplet's fall determines the probability that the droplet will reach the surface before evaporating. Droplets with radii smaller than 75 μm are unlikely to reach the ground because they fall so slowly.

Since we see clouds because sunlight reflects off individual droplets, the relatively short fall distance of even large cloud droplets contributes to the perception of a sharp cloud boundary.

10. *In theory, how long would it take a large (50-μm radius) cloud droplet to hit the ground if it was falling from a cloud base at 2000 m?*

11. *How long would it take a large raindrop (2500-μm radius) to reach the ground if it was falling from a cloud base of 2000 m?*

In reality, it is unlikely that a droplet with a 50-μm radius would make it to the surface without evaporating. Table 7-2 lists the maximum fall distance of various droplets before evaporation. You can see from the table why a large cloud droplet will evaporate before it reaches the ground.

12. *Approximately how far could drizzle-sized droplets fall before evaporating?*

Table 7-2

Drop radius (μm)	Maximum fall distance before evaporation (m)
2500	280,000
1000	42,000
500	1,000
200	500
100	150
50	0.1
10	0.033
2	0.00002
1	0.0000033

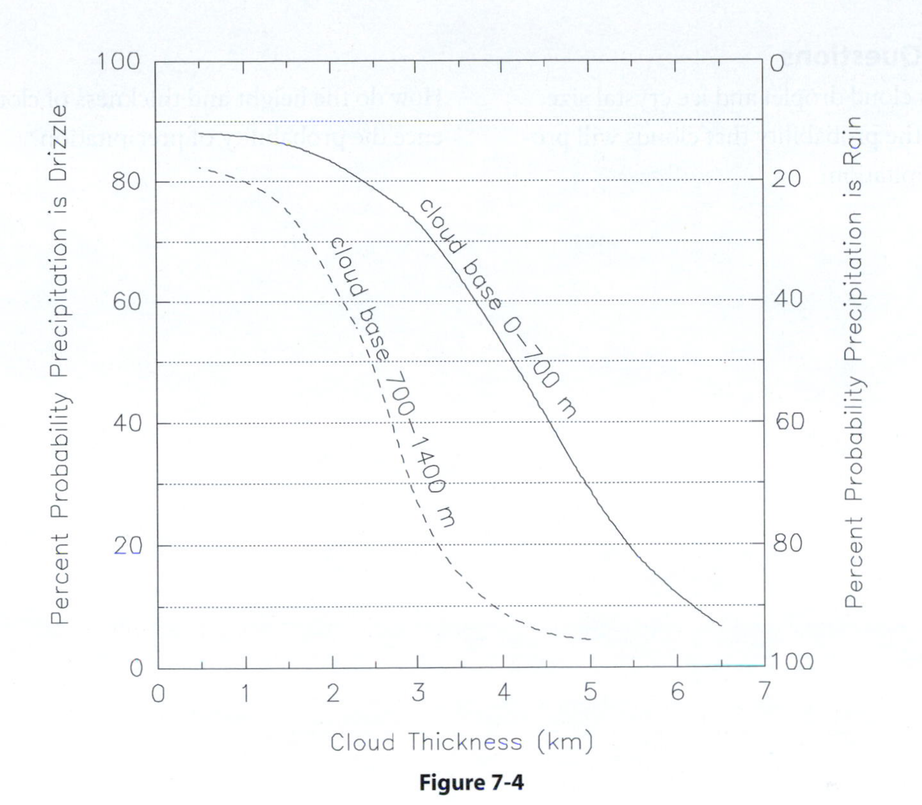

Figure 7-4

Figure 7-4 shows how cloud base height influences the probability that precipitation (if occurring) will reach the ground as drizzle or rain. The solid curve shows the probabilities for a low cloud base; the dashed curve shows the probabilities for a relatively high cloud base.

13. *If precipitation falls from a cloud 1 km thick with a low base, what is the probability that it will be drizzle?* __________ %
What is the probability that it will be rain?
__________ %

14. *If precipitation falls from a 5.5-km-thick cloud with a low base, what is the probability that it will be drizzle?* __________ %
What is the probability that it will be rain?
__________ %

15. *Why does the probability that the precipitation will be rain and not drizzle increase with a thicker cloud?*

16. a. *Consider precipitation from a cloud that is 3 km thick. If the cloud has a base below 700 m, what is the probability that the precipitation will be drizzle?* __________ %
What is the probability that the precipitation will be rain? __________ %

b. *If the cloud base is 700 to 1400 m high, what is the probability that the precipitation will be drizzle?* __________ %
What is the probability that the precipitation will be rain? __________ %

17. *Explain why precipitation from the higher cloud base is less likely to strike the ground as drizzle than precipitation from the lower cloud base.*

Review Questions

Why does cloud droplet and ice crystal size influence the probability that clouds will produce precipitation?

How do the height and thickness of clouds influence the probability of precipitation?

Lab 8

Atmospheric Motion

Materials Needed

- ruler

Introduction

What causes the wind to blow? This lab considers how pressure gradient, Coriolis, and frictional forces act to influence wind speed and direction. We examine how surface winds differ from upper-air winds and consider how horizontal winds lead to vertical motion.

Pressure Gradient Force

Horizontal winds are driven by horizontal differences in pressure between two locations. This pressure difference is referred to as a *horizontal pressure gradient*. The greater its value, the stronger the wind. Consider stations A and B, separated by 250 km.* The pressure is 1016 mb at station A and 1020 mb at station B—a 4-mb pressure difference over 250 km. We can express this pressure gradient as:

$$\frac{\Delta p}{d} = \frac{4\ mb}{250{,}000\ m}$$

The direction of pressure gradient force is from higher to lower pressure.

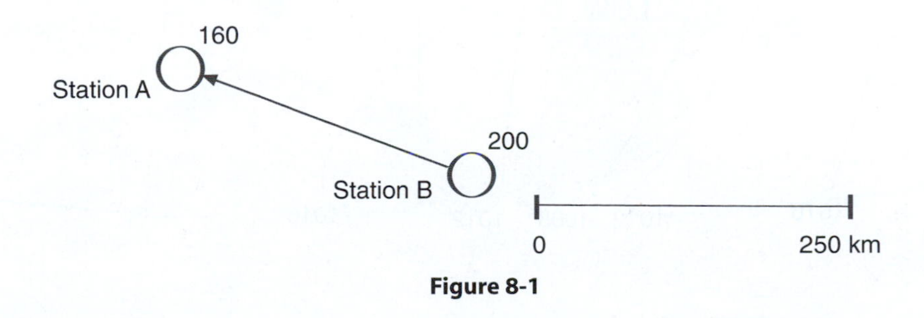

Figure 8-1

*Note how pressure is depicted on surface weather maps. To decode the values shown, first move the decimal one place to the left (e.g., station A with a value of 160 becomes 16.0). Then preface the number with a "9" or a "10". Since sea level pressure on earth is generally between 950 and 1050 mb, we can safely assume a pressure of 1016.0 mb at station A and 1020.0 mb at station B.

Surface weather maps include *isobars*—lines of constant pressure—that help us see horizontal pressure gradients. The U.S. map below shows isobars drawn at 4-mb intervals.

1. *Using Figure 8-2:*
   a. Circle the area with the greatest pressure gradient.
   b. Use arrows to show the direction of pressure gradient force at a few locations. (These are typically drawn perpendicular to isobars.)
   c. Label a region where you would expect the lightest winds.

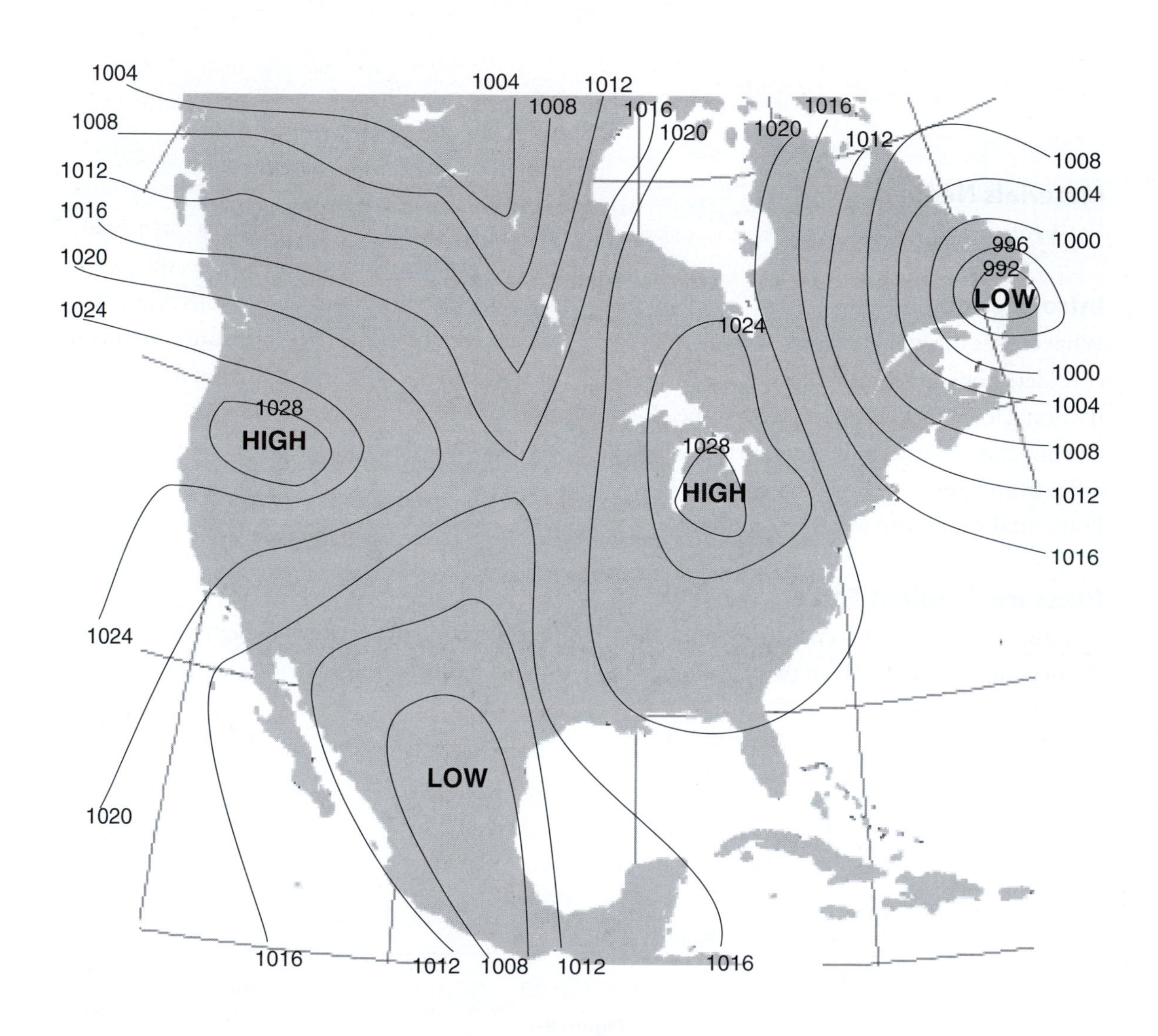

Figure 8-2

Optional Exercise: Mathematical Treatment of Pressure Gradient Force

The magnitude of the pressure gradient force is a function of the pressure difference between two points and air density. It can be expressed as:

$$\frac{F_{PG}}{m} = -\frac{1}{\rho} \bullet \frac{\Delta p}{d}$$

where F_{PG} = pressure gradient force

m = unit mass (1 kg)

ρ = density (kg m^{-3})

Δp = pressure change (pascals or kg m^{-1} s^{-2})

d = distance (m)

Let's consider an example where the pressure 5 km above Little Rock, Arkansas, is 540 mb and 5 km above St. Louis, Missouri, it is 530 mb. The distance between the two cities is 450 km, and the air density at 5 km is approximately 0.75 kg m^{-3}. In order to use the pressure gradient equation, we must use compatible units. We first convert pressure from millibars to pascals (Pa), another measure of pressure that has units of kilograms per meter per second squared. (Note that 1 mb = 100 Pa.) In our example the pressure difference above the two cities is 10 mb or 1000 Pa (1000 kg m^{-1} s^{-2}). Thus we have:

$$\frac{F_{PG}}{m} = -\frac{1}{0.75 \text{ kg m}^{-3}} \bullet \frac{1{,}000 \text{ kg m}^{-1} \text{ s}^{-2}}{450{,}000 \text{ m}} = -0.00296 \text{ m s}^{-2}$$

Newton's second law states that force equals mass times acceleration ($F = m \bullet a$).
In our example we have considered pressure gradient force per unit mass; therefore our result is an acceleration ($F \div m = a$). Because of the small units shown above, pressure gradient acceleration is often expressed as centimeters per second squared. In this example we have 0.296 cm sec^{-2}.

2. *Assuming a sea-level air density of 1.2 kg m^{-3}, calculate the magnitude of the pressure gradient force per unit mass in Figure 8-1.*

Coriolis Force

Most of our knowledge about the forces influencing wind is derived from Newton's laws of motion. These laws, however, are valid only when viewed from a fixed frame of reference. Since the earth rotates on its axis, our view of air movement is not from a fixed reference frame and we must compensate by considering the Coriolis force.

To understand why the Coriolis force causes a deflection from the path of motion we consider a simple drawing exercise.

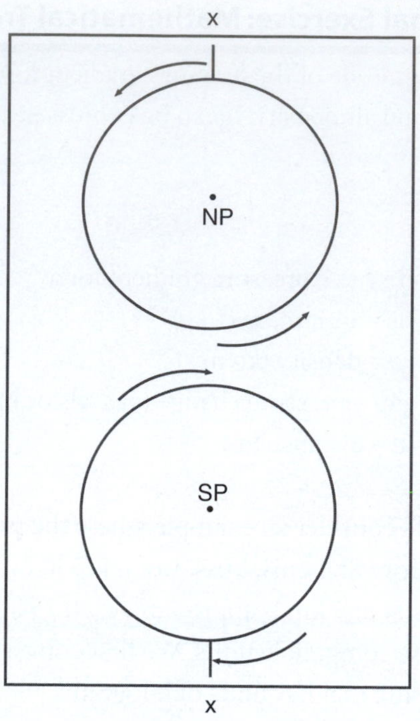

Figure 8-3

3. *For the Northern Hemisphere:*
   a. *Put an object at the top of your desk and align Figure 8-4 of your lab manual as in Figure 8-3, where x indicates the object.*

   b. *Draw a dashed line from the North Pole in Figure 8-4 directly toward the object at the top of the desk.*

   c. *Repeat step b, but this time draw a solid line toward the object at the top of the desk while you simultaneously rotate your lab book in the direction of the earth's rotation shown in Figure 8-4. (It is best to have someone turn the book while you draw the line.)*

4. *For the Southern Hemisphere:*
   a. *Put an object at the bottom of your desk and align page 73 of your lab manual as in Figure 8-3.*

   b. *Draw a dashed line from the South Pole in Figure 8-4 directly toward the object at the bottom of the desk.*

   c. *Repeat step b, but this time draw a solid line toward the object at the bottom of the desk while you simultaneously rotate your lab book in the direction of the earth's rotation shown in Figure 8-4.*

5. *How did the path of the solid line differ from the dashed line in the Northern Hemisphere example? What does this say about deflection relative to air motion in the Northern Hemisphere?*

6. *Describe the deflection relative to the path of motion in the Southern Hemisphere.*

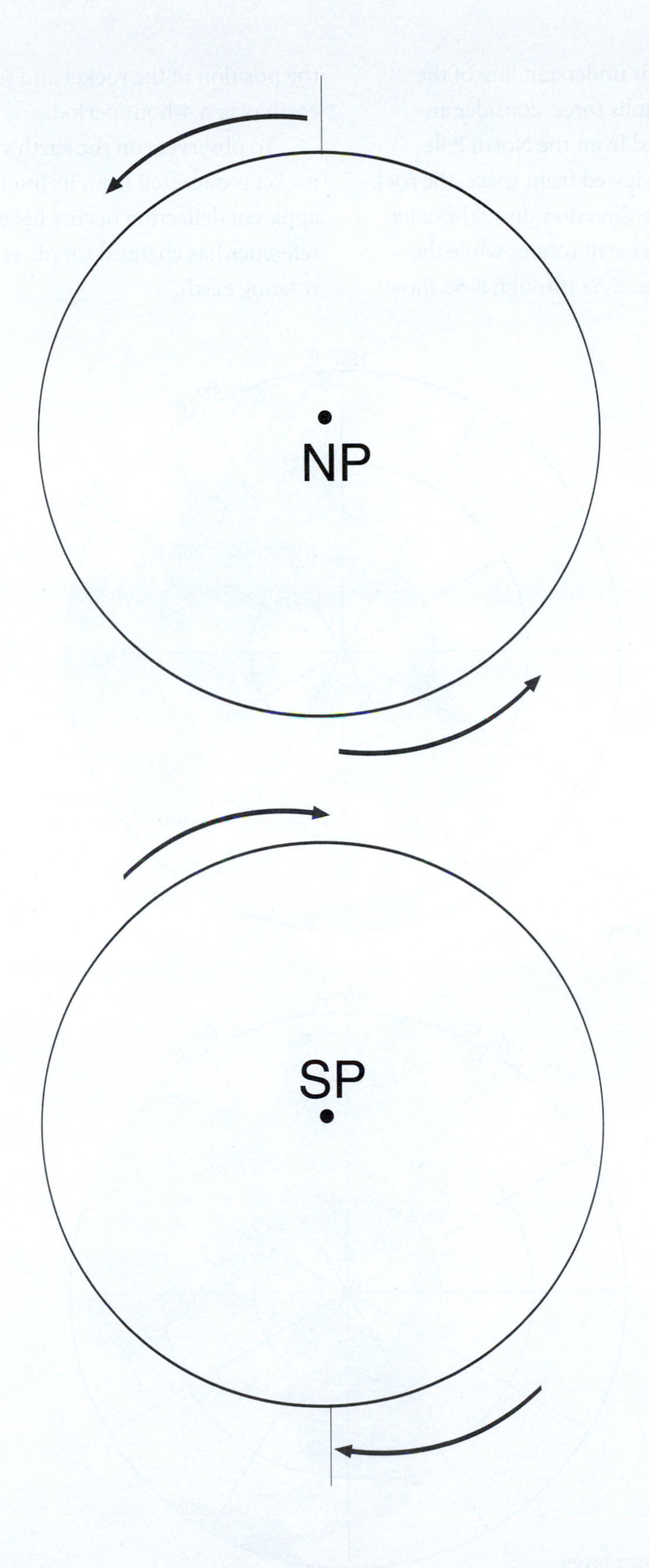

Figure 8-4

To reinforce your understanding of the direction of the Coriolis force, consider an unguided rocket fired from the North Pole toward Point A. As viewed from space, the rocket travels in the same direction throughout its flight. Of course, the earth rotates while the rocket travels. Figures 8-5a through 8-5c show the position of the rocket and rotation of the earth over a 4-hour period.

To observers on the earth's surface, the rocket is deflected from its intended path. This apparent deflection occurs because the frame of reference has changed for observers on the rotating earth.

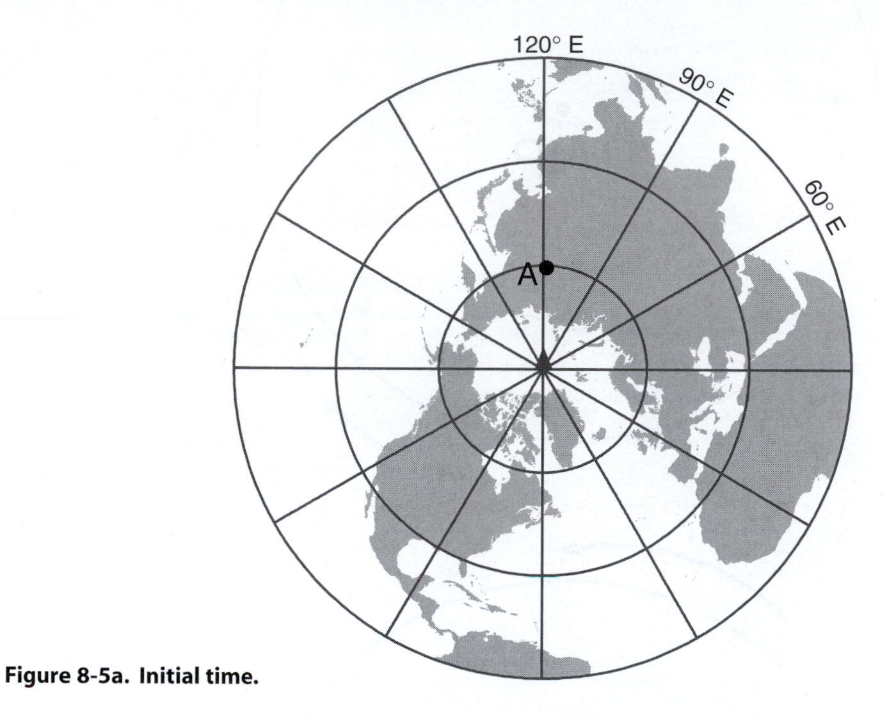

Figure 8-5a. Initial time.

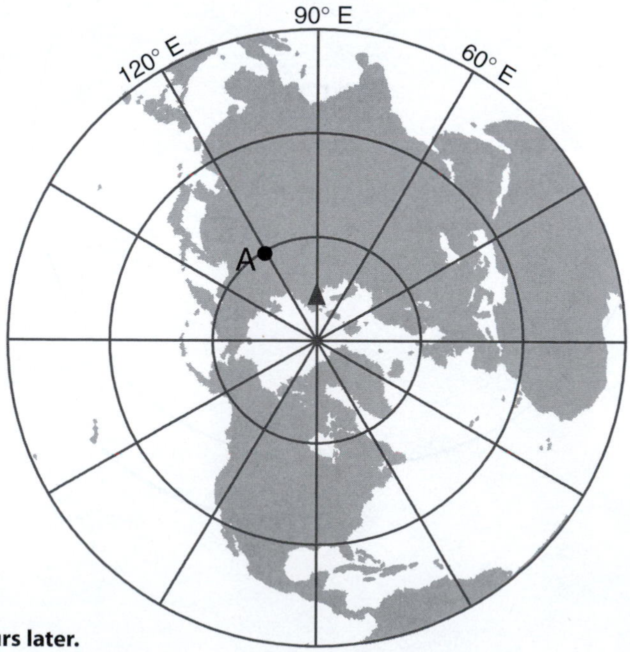

Figure 8-5b. Two hours later.

7. *Combine the position of the rocket at all three times onto Figure 8-5d by using the latitude and longitude information from the previous three figures. This should show the relative deflection from the perspective of someone on the earth's surface.*

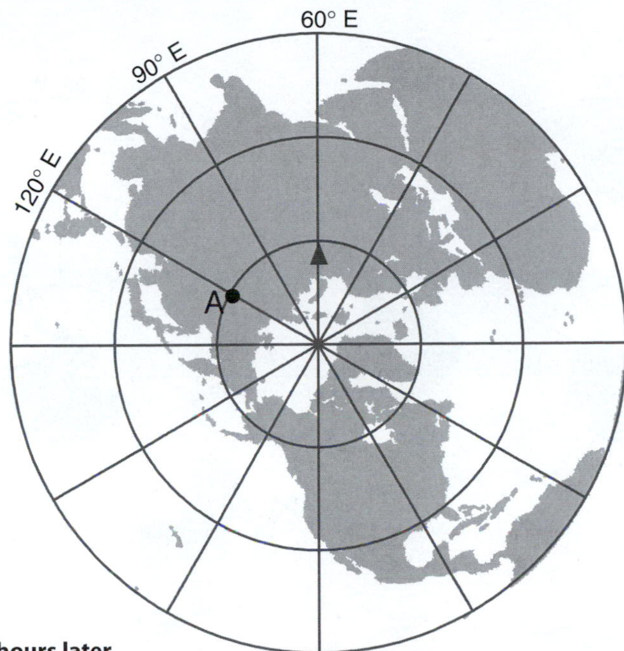

Figure 8-5c. Four hours later.

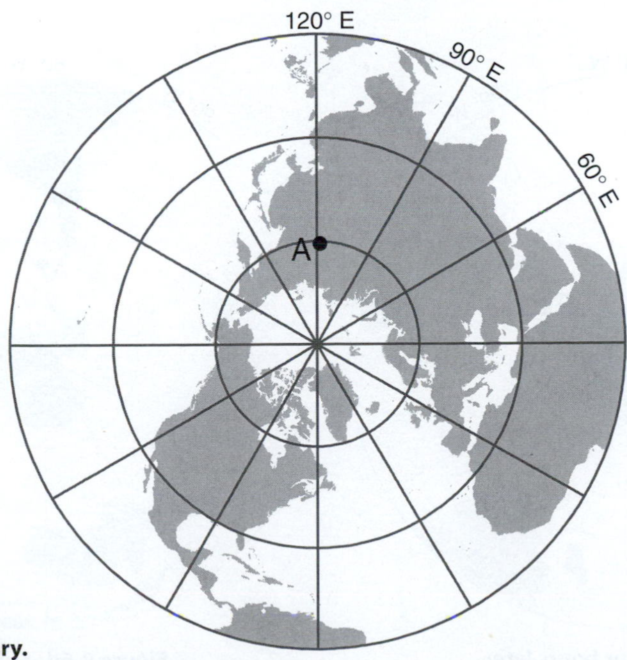

Figure 8-5d. Summary.

8. *Now show the apparent deflection in the Southern Hemisphere of a rocket fired from the South Pole. In Figure 8-6d, draw the position of the rocket at each of the three times depicted in Figures 8-6a through 8-6c.*

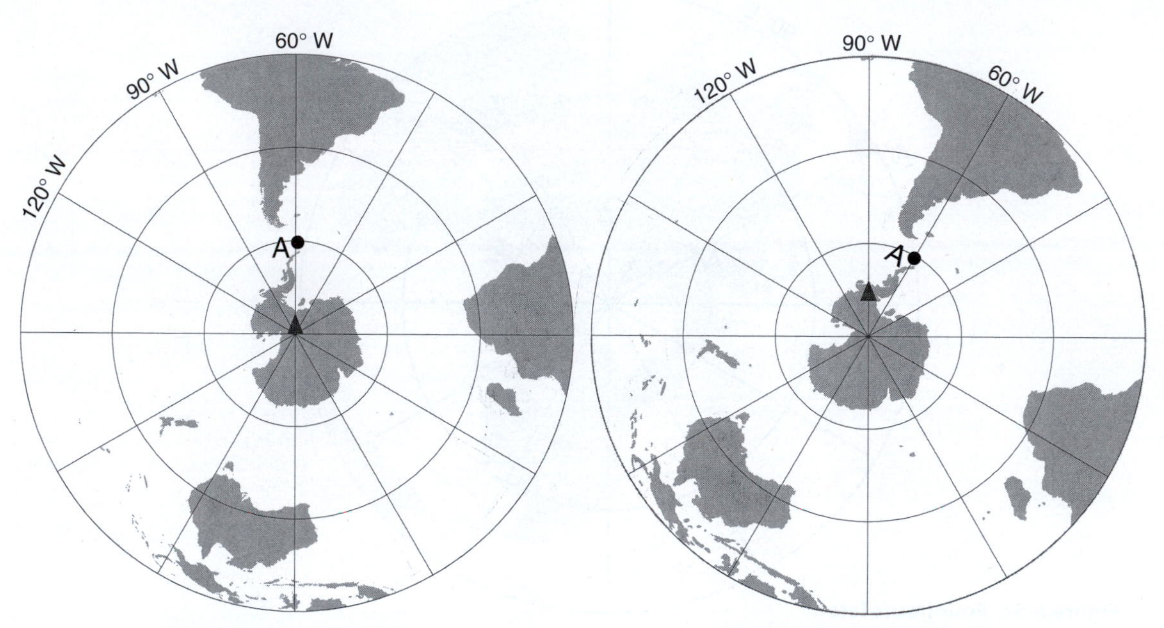

Figure 8-6a. Initial time.

Figure 8-6b. Two hours later.

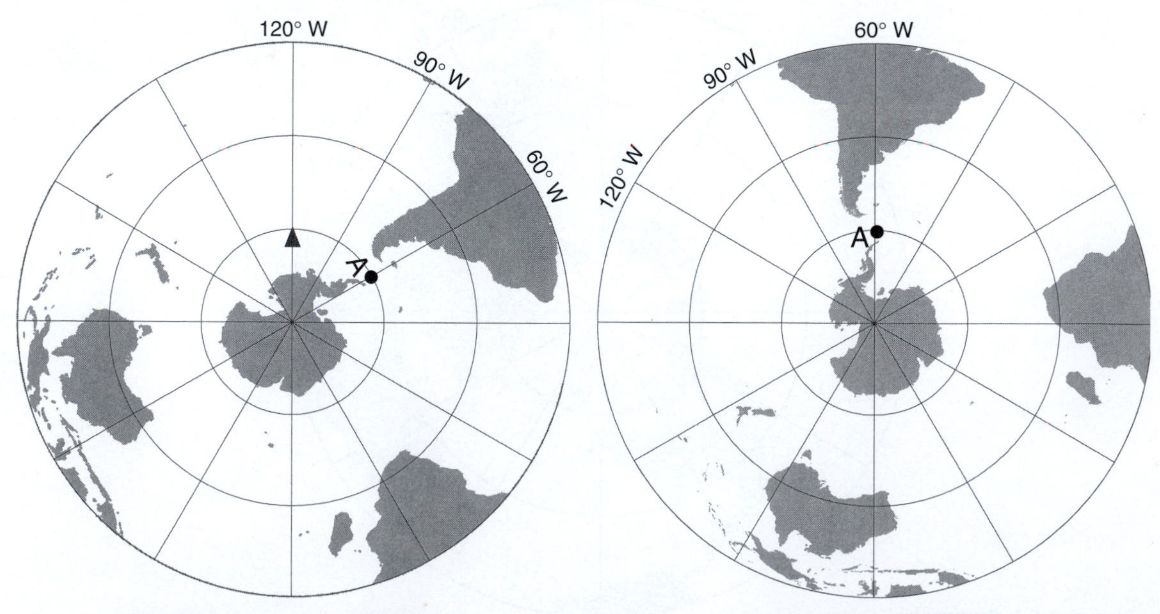

Figure 8-6c. Four hours later.

Figure 8-6d. Summary.

The Coriolis force varies with wind speed. Imagine in Figure 8-7 that the rocket travels twice the speed of the previous example (and therefore covers twice the distance in 4 hours). Its intended target is Point B.

9. *Notice that the rocket travels twice as fast as in the previous example. Now combine the position of the rocket at all three times onto Figure 8-7d by using the latitude and longitude information from the previous three figures.*

10. *Is the rocket in Figure 8-7d displaced from its intended target a greater or lesser distance than in Figure 8-5d? Extrapolating from this, does Coriolis force increase or decrease with wind speed?*

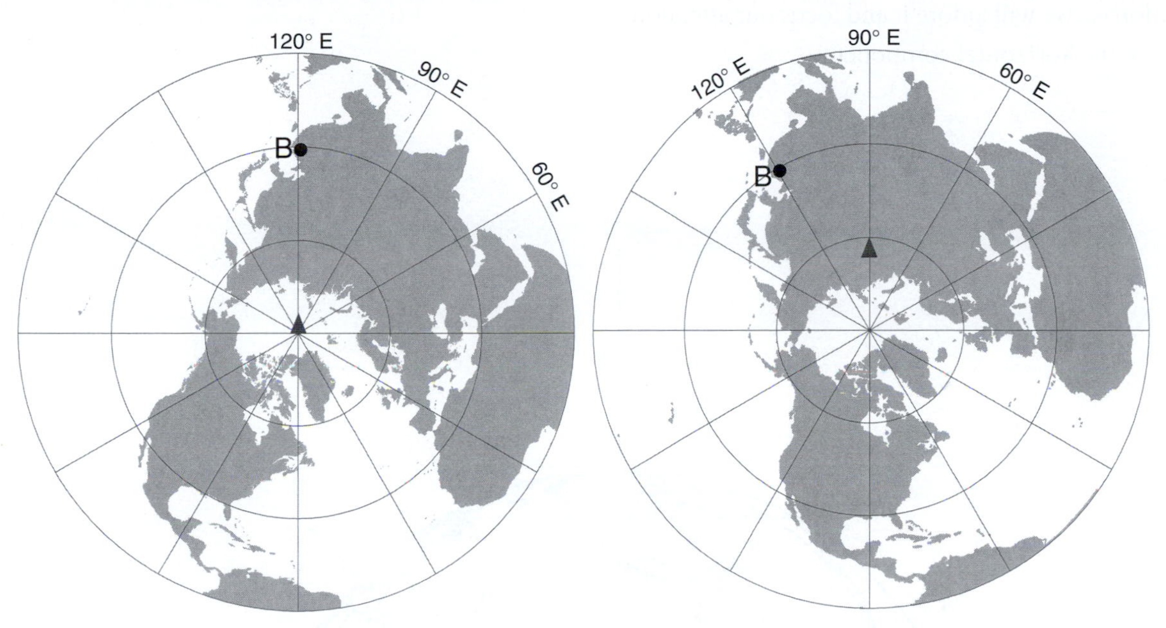

Figure 8-7a. Initial time.

Figure 8-7b. Two hours later.

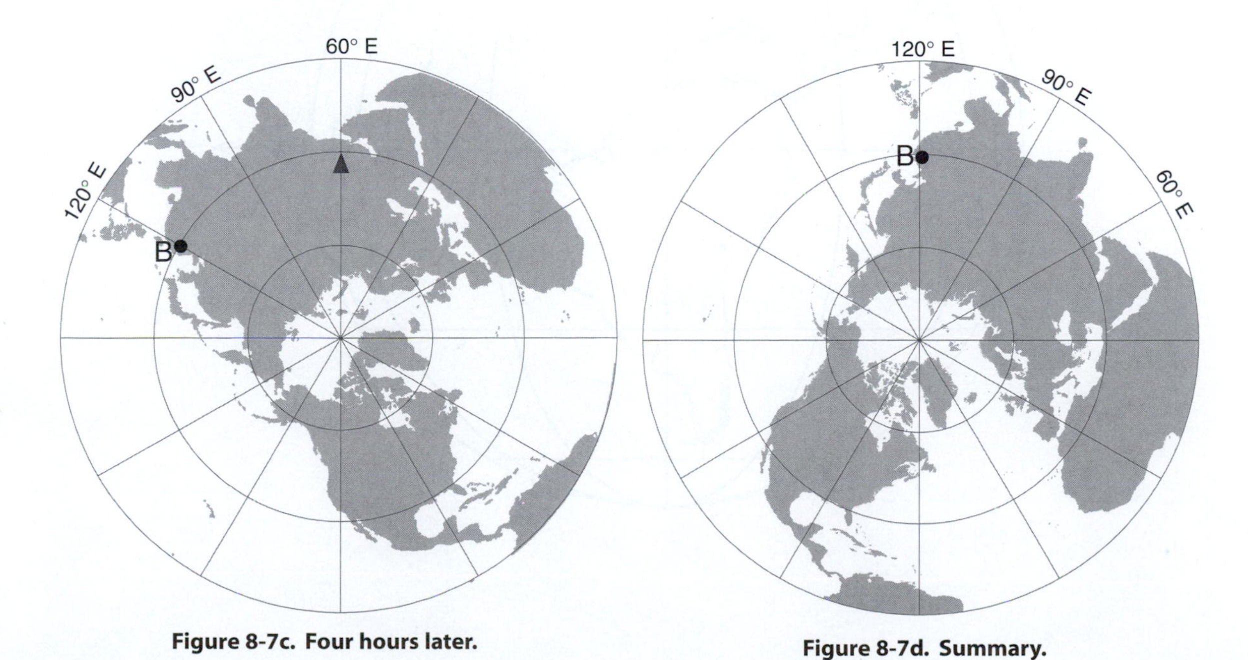

Figure 8-7c. Four hours later.

Figure 8-7d. Summary.

Since the Coriolis force results from the earth's rotation, it is expressed perpendicular to the earth's axis (Figure 8-8). Therefore, at the poles it is completely horizontal and at the equator it is completely vertical. At other latitudes, it has both horizontal and vertical components (shown by the dashed lines at 40° W latitude). Because the vertical component of the Coriolis force is small compared with other vertical forces, we will ignore it and focus our attention on the horizontal component.

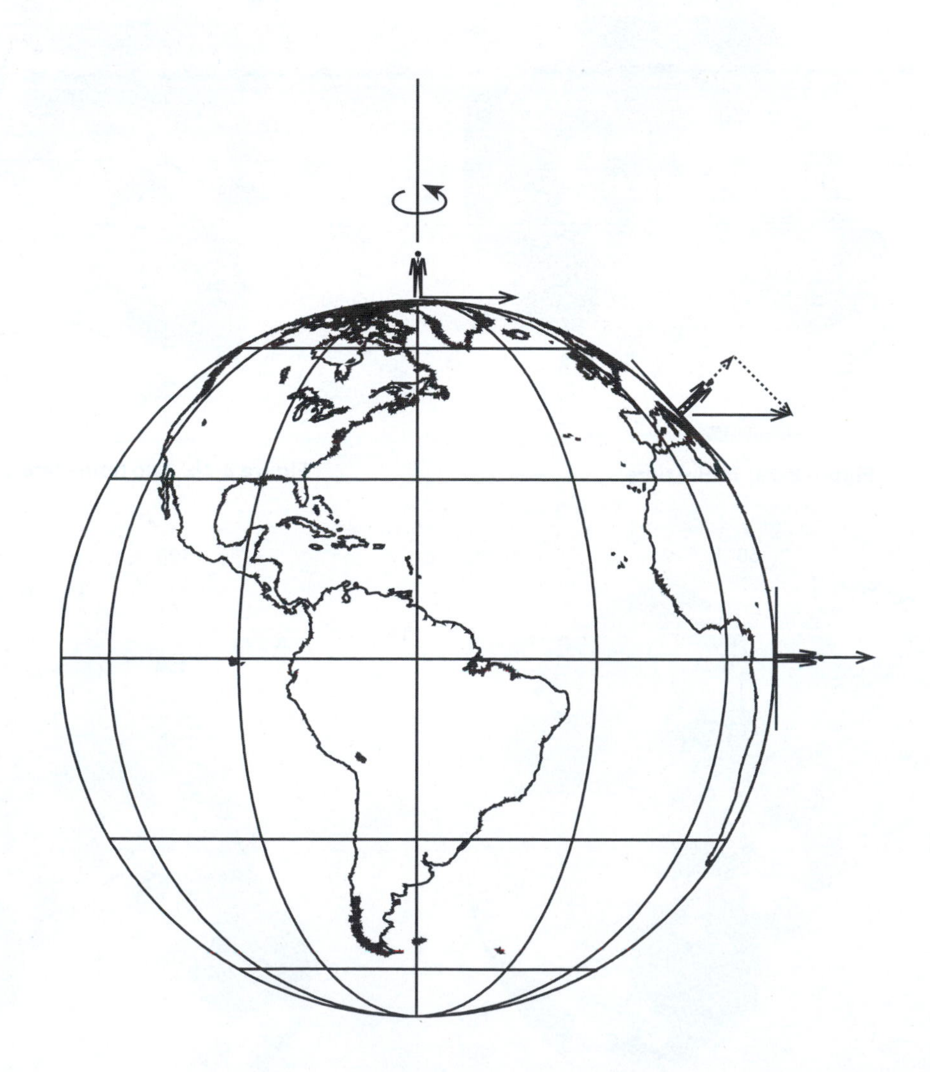

11. *The Coriolis force is depicted at two different latitudes in Figure 8-9. Draw the vertical and horizontal component of the Coriolis force at the two locations. Is the horizontal component greater at the higher latitude or the lower latitude?*

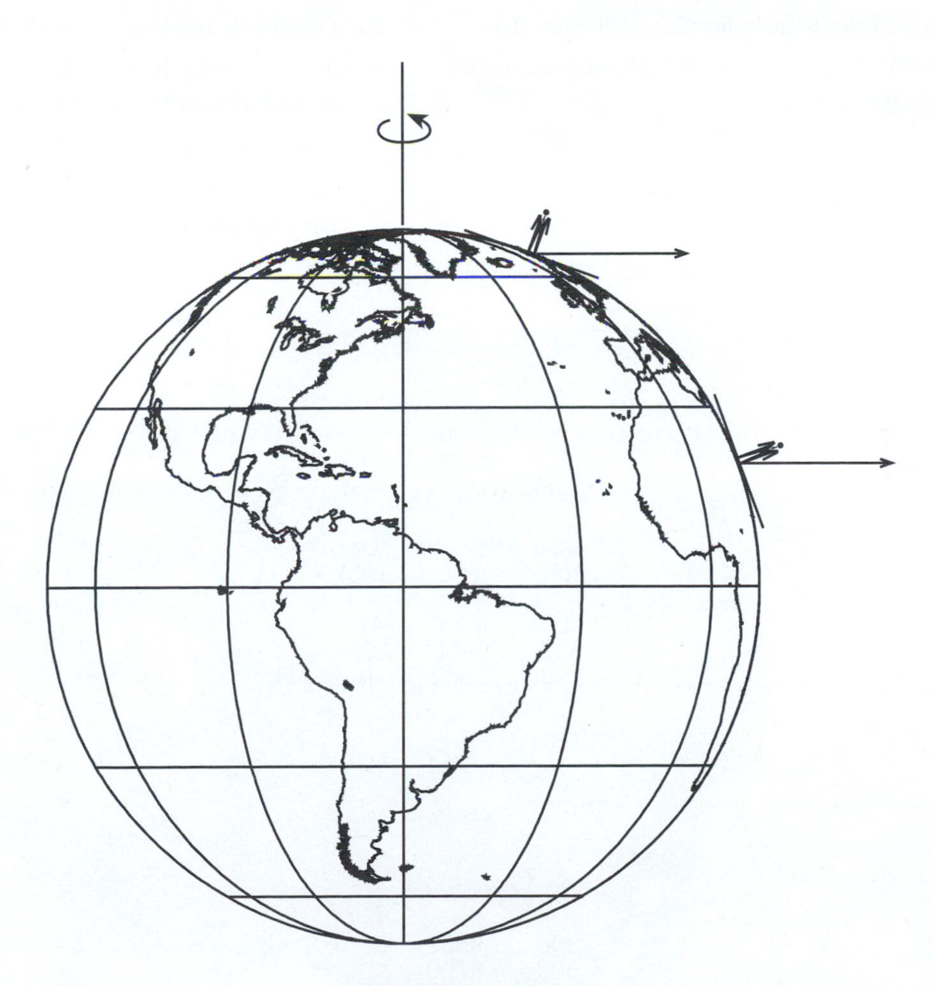

Figure 8-9

Upper-Level Winds

From Lab 1, you will remember that pressure decreases with height above the earth's surface. We express this decrease in a manner similar to the horizontal pressure gradient:

$$\frac{\Delta p}{\Delta z}$$

where Δp = pressure change
and Δz = height change

If a pressure gradient force is directed upward, it acts to move air vertically. Luckily for creatures breathing at the earth's surface, this force is counteracted by gravity, creating a balance we refer to as *hydrostatic equilibrium*. The hydrostatic equation shows this balance mathematically:

$$\frac{\Delta p}{\Delta z} = -\rho g$$

where

$\frac{\Delta p}{\Delta z}$ = the upward-directed pressure gradient force

ρ = density (kg m^{-3})

and g = gravity (9.8 m s^{-2})

This relationship is relevant to how we depict winds on upper-air maps.

12. *Explain why pressure decreases more rapidly with height near the surface than at higher altitudes.*

13. *Consider the two locations on the North American map in Figure 8-10. Following the example for 850 mb, plot the heights of the 500- and 300-mb pressure levels in Figure 8-11. Draw lines connecting the two points for each pressure level; label each line you draw with the appropriate pressure value.*

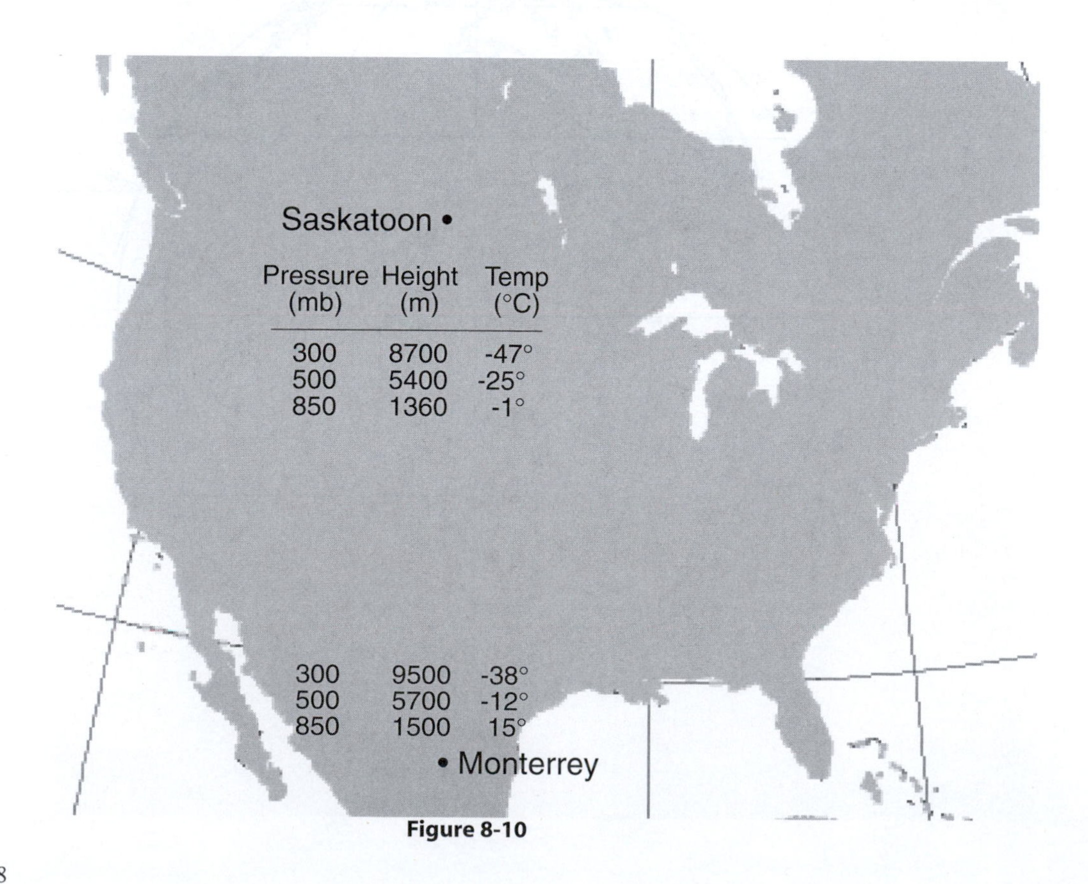

Figure 8-10

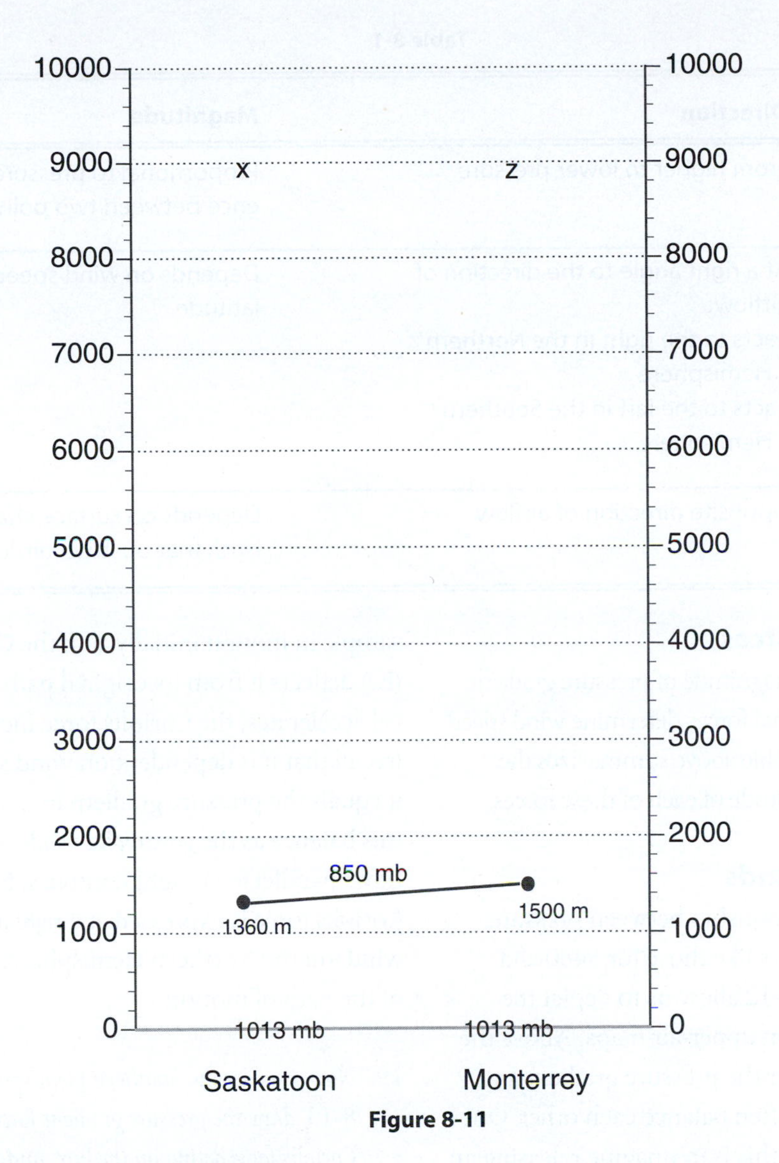

Figure 8-11

14. *At which location, Saskatoon or Monterrey, does pressure decrease with height most rapidly?*

15. *How does this relate to temperature differences between the two sites?*

16. *Both points x and z are at 9000 m. Which point has the highest pressure? Draw arrows between the points showing the direction of the horizontal pressure gradient force at 9000 m.*

17. *At what height, 1500 m or 9000 m, is the horizontal pressure gradient greater?*

18. *How should wind speeds at these two heights reflect this difference?*

Table 8-1

Force	Direction	Magnitude
Pressure gradient	From higher to lower pressure	Proportional to pressure difference between two points
Coriolis	At a right angle to the direction of airflow: • acts to the right in the Northern Hemisphere • acts to the left in the Southern Hemisphere	Depends on wind speed and latitude
Friction	Opposite direction of airflow	Depends on surface characteristics that obstruct airflow

Summary of Forces

The direction and magnitude of pressure gradient, Coriolis, and frictional forces determine wind speed and direction. The table above summarizes the direction and magnitude of each of these forces.

Geostrophic Winds

Because of the relationship between pressure and height, contours like those for 5460 and 5520 m in Figure 8-12 allow us to depict the pressure gradient on upper-air maps. Above the earth's friction layer the pressure gradient and the Coriolis force often balance each other. One way to understand this is to imagine releasing an air sample like that shown by the box in Figure 8-12. The pressure gradient force (F_{pg}) sets the sample in motion initially, and the Coriolis force (F_{co}) deflects it from its original path. As the parcel accelerates, the Coriolis force increases (recall that it is dependent on wind speed), until it equals the pressure gradient force. We define this balance as the *geostrophic wind*—one that flows parallel to straight contours. Note that the Coriolis force is expressed at a *right* angle to the wind. (In the Northern Hemisphere, to the right of the path of motion.)

19. *Now consider the Southern Hemisphere. In Figure 8-13, draw the pressure gradient force and the Coriolis force acting on the box, and the resulting geostrophic wind.*

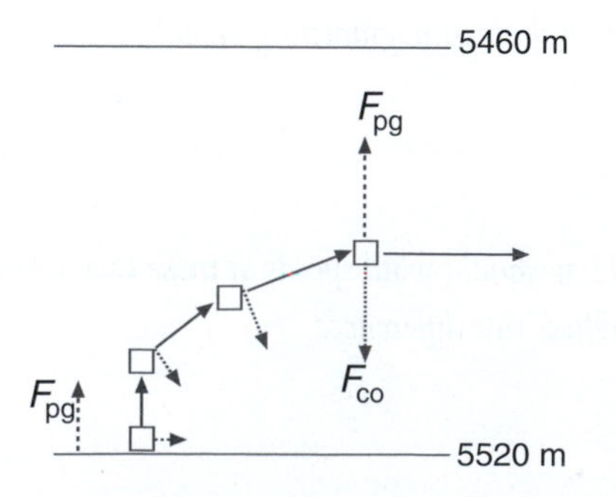

Figure 8-12. 500-mb geostrophic wind in the Northern Hemisphere.

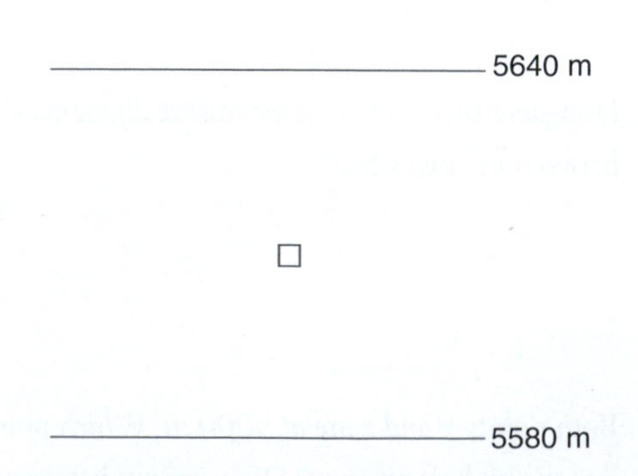

Figure 8-13. 500-mb geostrophic wind in the Southern Hemisphere.

The Gradient Wind

In reality, contours are typically curved as in Figure 8-14, and upper-level winds often flow parallel to them. We define a *gradient wind* as one that blows parallel to the contours. Unlike the geostrophic wind, the gradient wind accelerates (since it is changing direction). The acceleration occurs because the pressure gradient and Coriolis forces no longer balance each other.

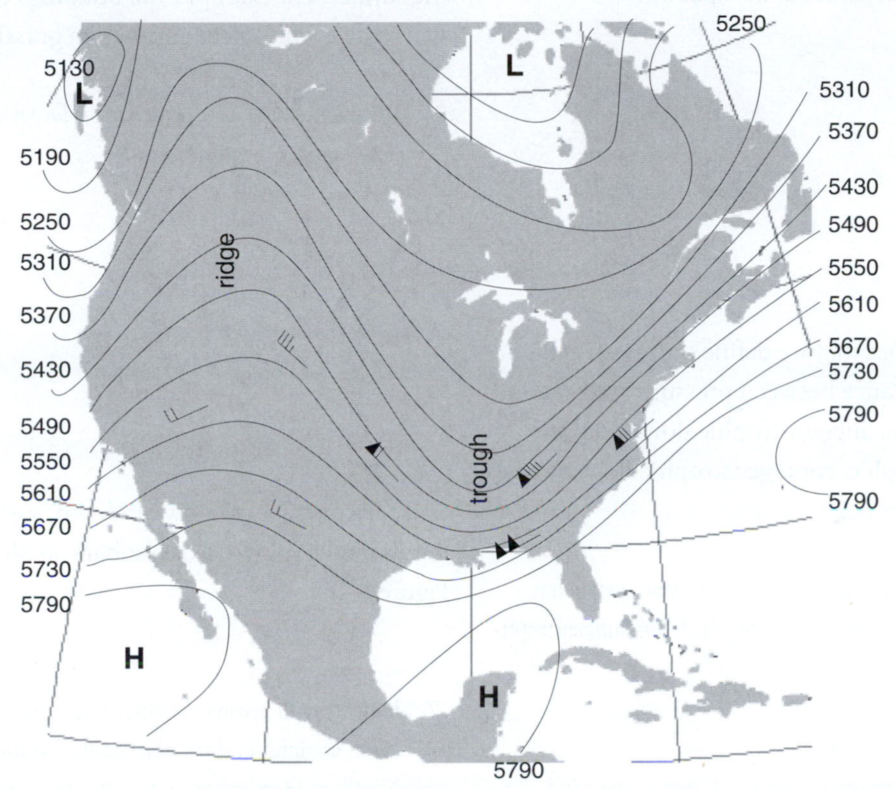

Figure 8-14

20. *In the figures below, draw arrows showing the pressure gradient and Coriolis force acting on each box. Since the pressure gradient is the same in both drawings, you should make the pressure gradient force vector the same for all six boxes. However, you should vary the size of the Coriolis vector in a way that would maintain the wind in a curved path.*

Since the gradient wind changes its direction, we say that it is accelerating. Differences between the pressure gradient and Coriolis force cause this acceleration, often called the centripetal acceleration.

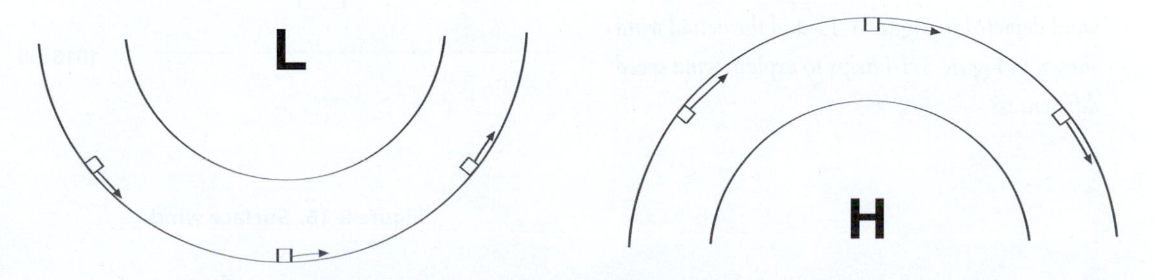

Figure 8-15. Gradient wind around Northern Hemisphere trough and ridge.

21. In Figure 8-15, the pressure gradient force around the low is the same as that around the high. However, there is a difference in the magnitude of Coriolis force around each. What does this say about the wind speed around the low versus the high in this particular example?

Remember that we define the geostrophic wind as a balance between pressure gradient and Coriolis force. Subgeostrophic flow is slower than geostrophic, supergeostrophic flow is faster than geostrophic.

22. In Figure 8-15, label which pressure system has supergeostrophic flow and which has subgeostrophic flow.

23. Now reexamine Figure 8-14. Notice the wind speed differences between the trough (low pressure) vs. the ridge (high pressure). Does this conform or contrast with your previous answer about subgeostrophic and supergeostrophic flow?

24. What difference between the theoretical gradient wind depicted in Figure 8-15 and the actual wind shown in Figure 8-14 helps to explain wind speed differences?

Surface Winds and Friction

Winds close to the earth's surface are also influenced by friction, which acts in a direction opposite the wind, slowing the wind. The magnitude of friction depends on the "roughness" of the surface. For example, tall buildings or forests will reduce wind speed more than grasslands.

25. What affect will a reduction in surface wind speed have on the Coriolis force?

Since friction acts to slow air down, surface winds tend to blow across isobars, as shown in Figure 8-16.

26. Indicate with arrows the three forces (pressure gradient, Coriolis, and friction) acting on the Northern Hemisphere box in the diagram below.

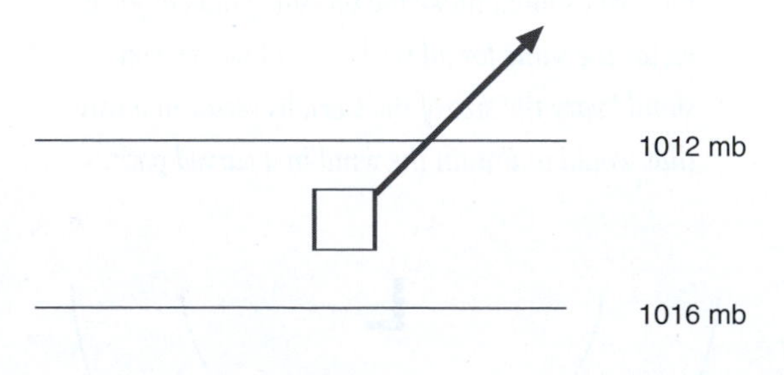

Figure 8-16. Surface wind.

27. *Figure 8-17 shows winds circulating around surface high and low pressure centers in both the Northern and Southern Hemispheres. Label the correct hemispheres (Northern or Southern) and pressure system (high or low) for each of the four examples.*

28. *Choose one box on each example and draw the pressure gradient, Coriolis, and frictional forces associated with it.*

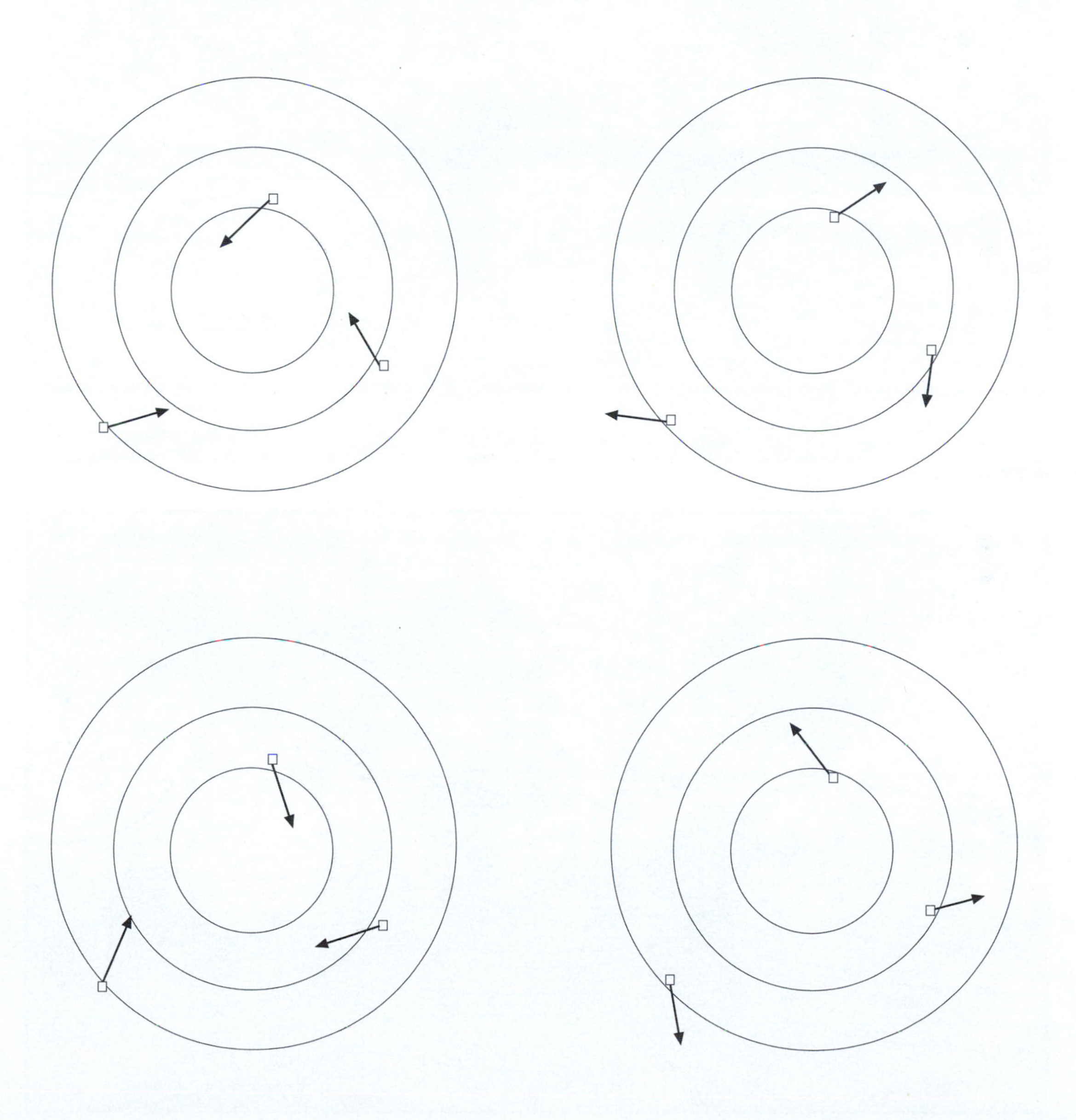

Figure 8-17

Surface Air Pressure and Associated Global Circulation

The forces producing the wind act at a variety of scales, from a few meters (microscale) to tens of thousands of meters (planetary scale). Note the January and July surface pressure and wind patterns in the maps below. As you can see, the circulation around high and low pressures at the global scale reflects the same patterns that you drew on the previous pages.

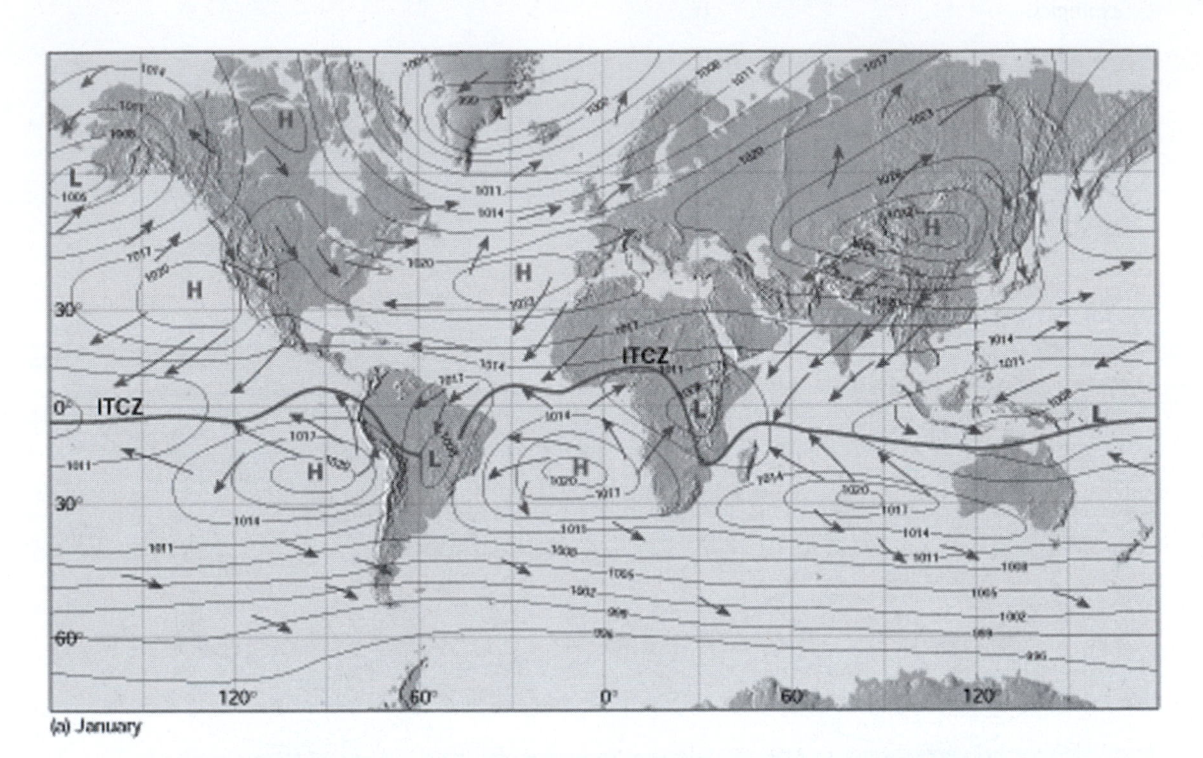

(a) January

(b) July

Figure 8-18. Surface pressure and wind patterns.

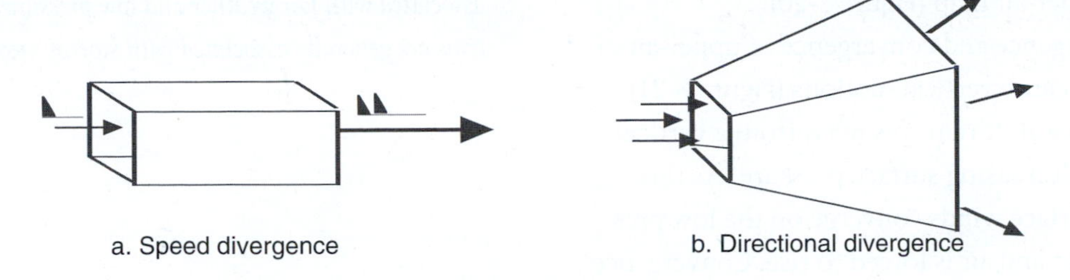

Figure 8-19

Winds and Vertical Motion

As a result of pressure gradient, Coriolis, and frictional forces, surface winds will converge toward low pressure centers. If a low pressure center is to be maintained, there should be a net outflow of air in a column extending from the surface to upper levels of the troposphere. In contrast, maintaining a surface high pressure center requires a net inflow of air to the column.

We can simplify the concept of net outflow by considering the volume of air entering and exiting a box in the upper troposphere. Two phenomena could contribute to a net outflow from the box: (1) speed divergence results from changes in wind speed between the entering and exiting regions of the box (Figure 8-19a), and (2) directional divergence (diffluence) results from changes in wind direction between the entering and exiting regions (Figure 8-19b).

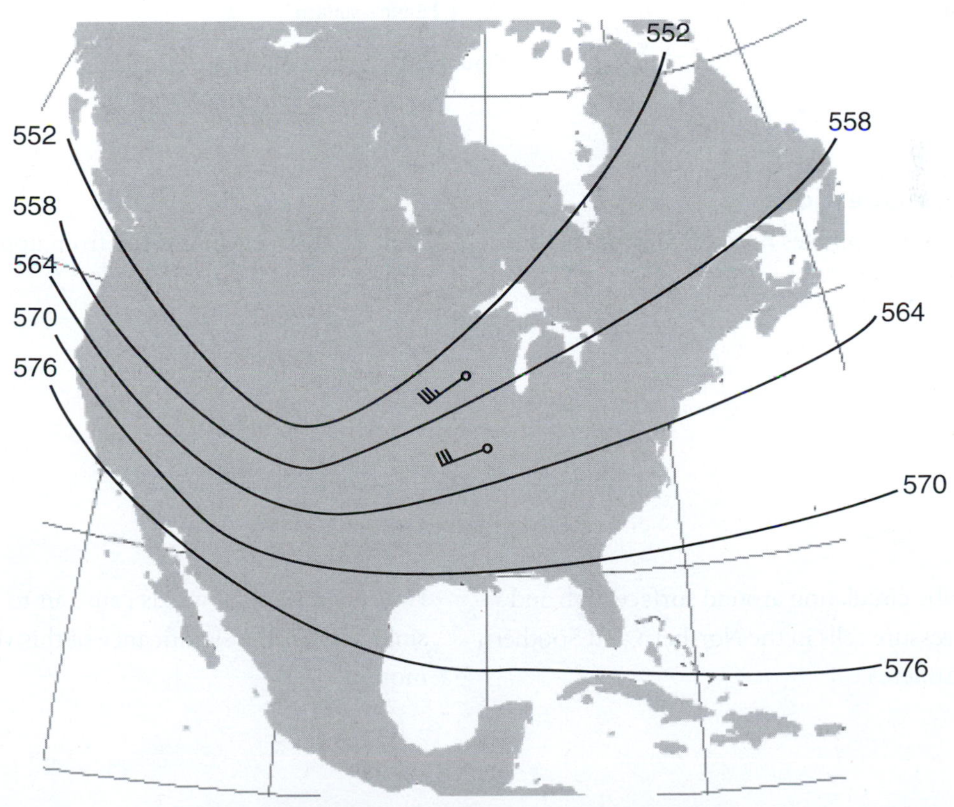

Figure 8-20. Directional divergence on a 850-mb map.

Directional divergence can also be viewed on an upper-air map (Figure 8-20).

Divergence and convergence of upper-air flow also cause vertical motions (Figure 8-21). Divergence aloft removes mass from a vertical column, decreasing surface pressure. As this occurs surface winds converge on the low pressure center and air is forced to rise. Convergence aloft adds mass to a vertical column of air and therefore increases pressure. When it occurs high in the troposphere, convergence causes sinking, as the inversion of the stratosphere restricts rising motion.

29. *Explain why high pressure systems are generally associated with fair weather and low pressure systems are generally associated with stormy weather.*

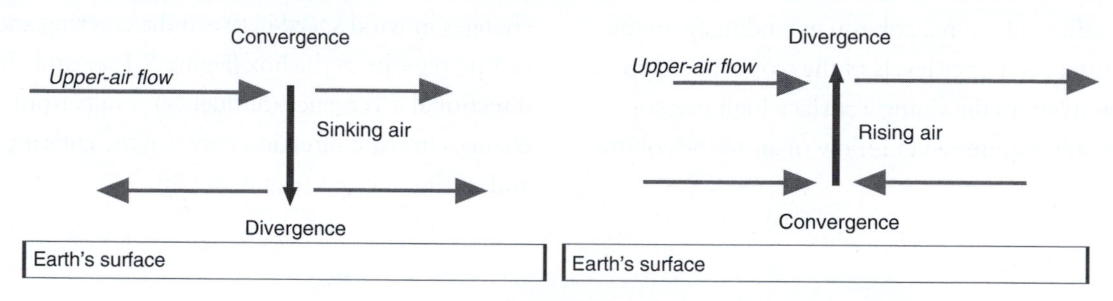

Figure 8-21

Review Questions

Which force initiates horizontal wind?

How do surface winds differ from upper-air winds?

Describe circulation around surface high and low pressure cells in the Northern and Southern Hemispheres.

How do horizontal winds cause air to rise or sink? What is the significance of this vertical motion?

Lab 9

Weather Map Analysis

Introduction

With a few exceptions (e.g., clouds), most atmospheric processes are invisible. How then do we "see" the weather in order to forecast its changes? The purpose of this lab is to learn how to construct and interpret weather maps. We will focus on the mid-latitudes, where identification of air masses, fronts, and mid-latitude cyclones can help meteorologists forecast changing weather patterns.

Surface Weather Maps

Every six hours atmospheric data are collected at approximately 10,000 surface weather stations around the world. These data are transmitted to one of three World Meteorological Centers, in Melbourne, Australia; Moscow, Russia; or Washington, D.C. Weather data are disseminated to national meteorological centers where *synoptic-scale* maps are generated. *Synoptic* means coincident in time, and a synoptic map is a map of weather conditions for a specific time. By convention, the time printed on many weather maps is Greenwich Mean Time (GMT, also called Coordinated Universal Time), the time at the prime meridian. Meteorologists often call this *Zulu* or *z* time. Thus, a map labeled 1200z shows conditions at noon in London, which is 7:00 AM EST in New York.

In the United States an automated weather network collects hourly surface data. Since each station collects data for as many as 18 weather characteristics, a compact method of symbolization must be used to include all this information on a single weather map. The station model, developed by the World Meteorological Organization, is the standard format for symbolizing weather characteristics. Figure 9-1 illustrates the arrangement of data in the WMO model; Appendix D provides a complete list of symbols used in this lab.

1. *Decode information from each of the following station models:*

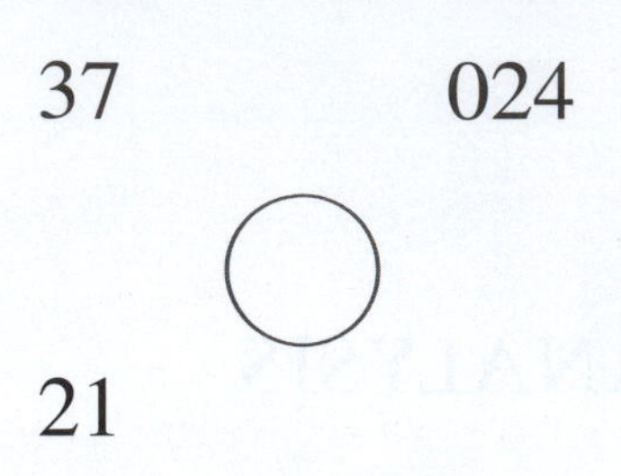

Barometric pressure __________
Temperature __________
Dew-point temperature __________

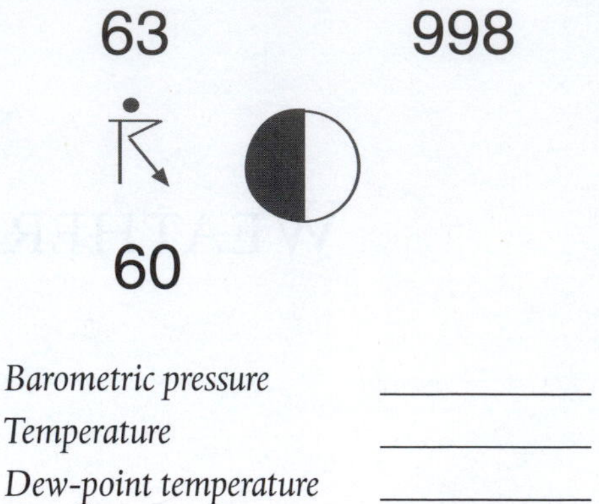

Barometric pressure __________
Temperature __________
Dew-point temperature __________
Sky coverage __________
Current weather __________

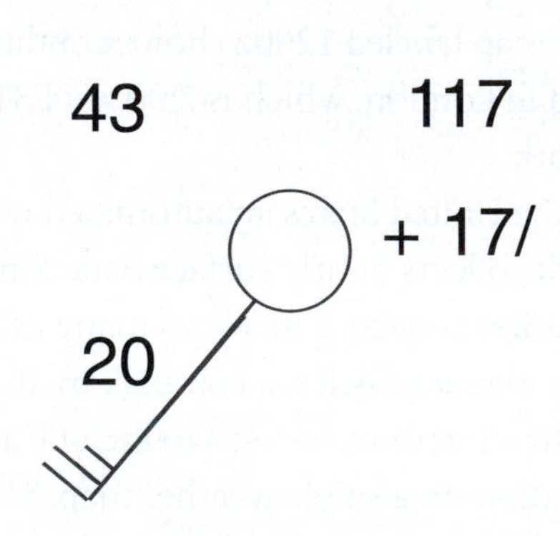

Barometric pressure __________
Temperature __________
Dew-point temperature __________
Sky coverage __________
Wind speed __________
Wind direction __________
Pressure change during last 3 hours __________
Pressure tendency __________

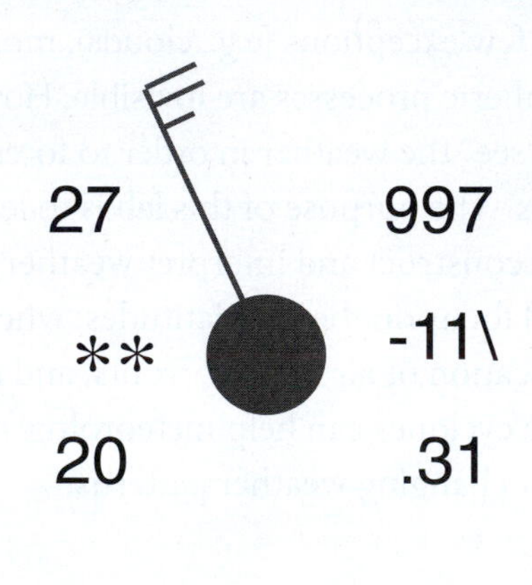

Barometric pressure __________
Temperature __________
Dew-point temperature __________
Sky coverage __________
Current weather __________
Wind speed __________
Wind direction __________
Pressure change during last 3 hours __________
Pressure tendency __________

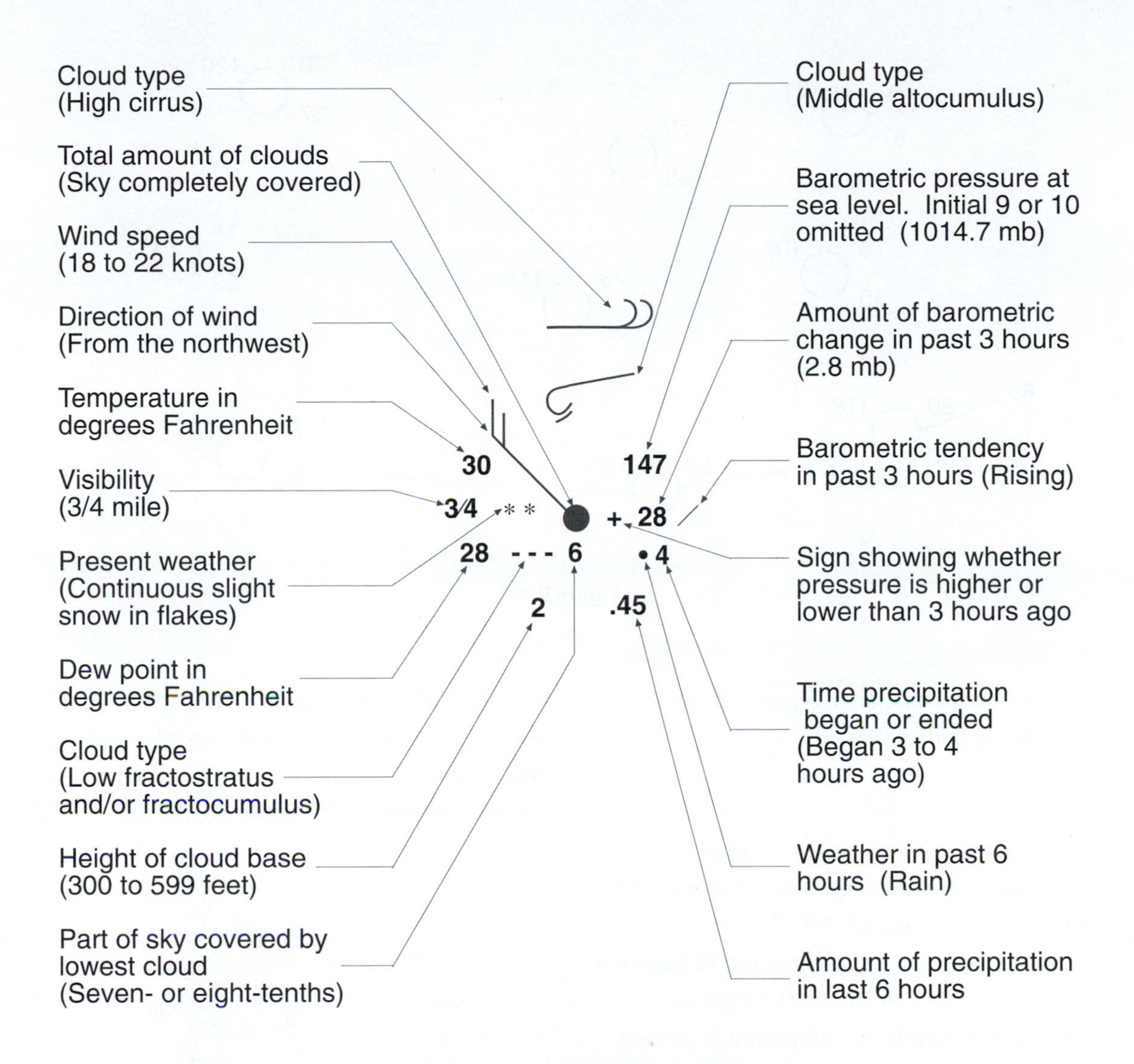

Figure 9-1. WMO station model.

Mapping Spatial Patterns of Meteorological Variables

Weather maps are most useful when their information is analyzed in some fashion. Highlighting the spatial patterns of specific variables—such as temperature, dew point, pressure, and winds—is a first step in weather analysis. We often use *isolines* (lines of constant value) for this purpose. Each type of isoline is named to reflect the variable being mapped: *isotherms* are lines of constant temperature; *isobars* are lines of constant barometric pressure; and *isodrosotherms* are lines of constant dew point.

Today meteorologists often use computer programs to draw isolines. Here we will construct some manually to better understand them. As a starting point, consider the isotherm drawn for 80°F in Figure 9-2.

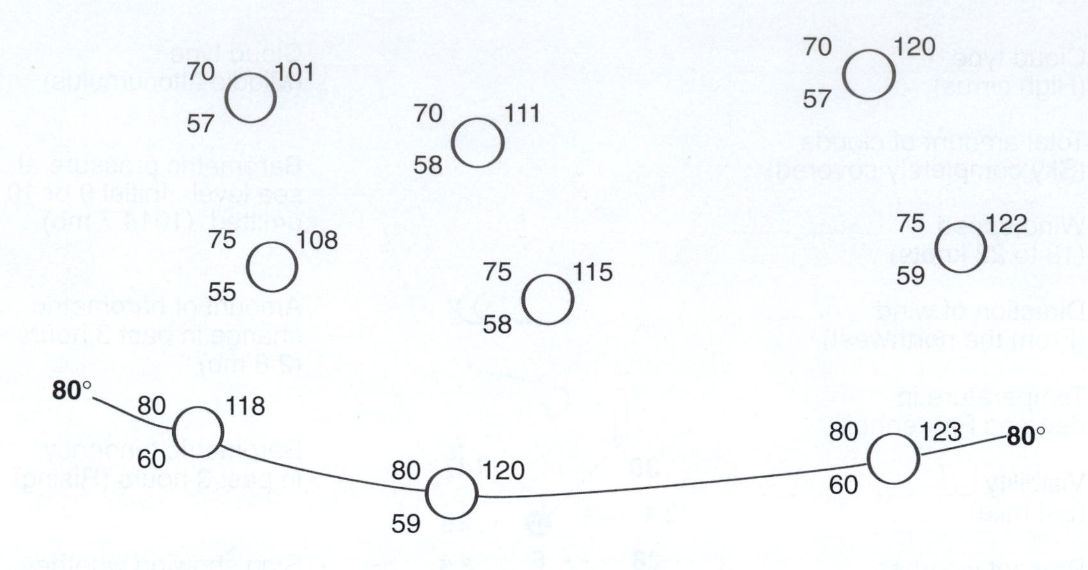

Figure 9-2

2. *Complete the analysis in Figure 9-2 by constructing isotherms at 70°F and 75°F.*

Drawing isolines in the previous example was straightforward, since the temperature at each station was exactly 80°, 75°, or 70°F. Because such patterns rarely occur in nature it is often necessary to *interpolate* between points. For example, we may want to draw a 1012-mb isobar (line of constant barometric pressure) using the following station data:

140
100
130
125

In a simple interpolation scheme we might decide that 1012 mb is exactly between 1010.0 and 1014.0 mb and would indicate this position with a small "x." A value of 1012 mb would also exist at two-thirds the distance between 1010.0 and 1013.0 mb, and at four-fifths the distance between 1010.0 and 1012.5 mb.

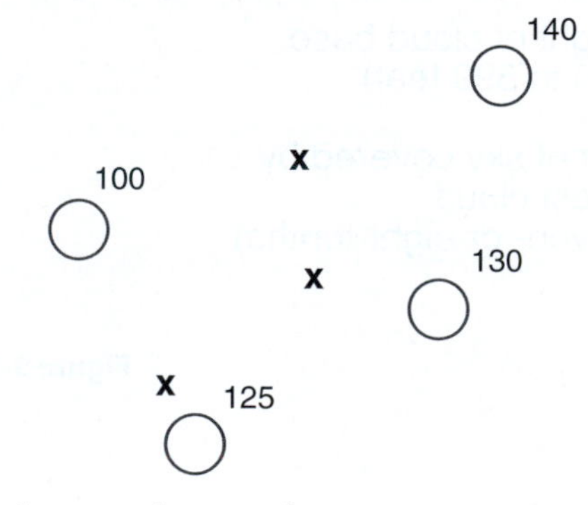

The "x's" that we draw represent new data points with a value of 1012 mb, through which we construct an isoline labeled "1012."

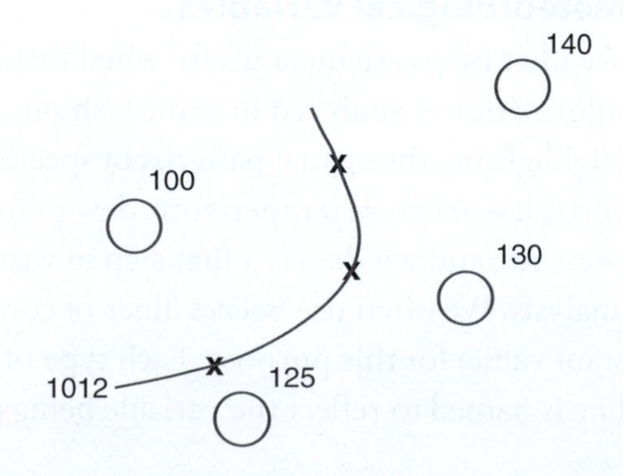

There are some conventions meteorologists use in constructing isolines. Study the isodrosotherms below.

- Because isolines are lines of constant value, they do not cross.

- Isolines should be relatively smooth. Sharp breaks are rare.

- They should be drawn at fixed intervals. Meteorologists traditionally use 4-mb intervals, centered on 1000 mb, for barometric pressure (e.g., 996 mb, 1000 mb, 1004 mb, etc.). For temperature and dew point, intervals of 5°F are commonly used.

- Isolines should be labeled near the edge of the map. When they form a closed figure, the label is inserted in a small break in the line.

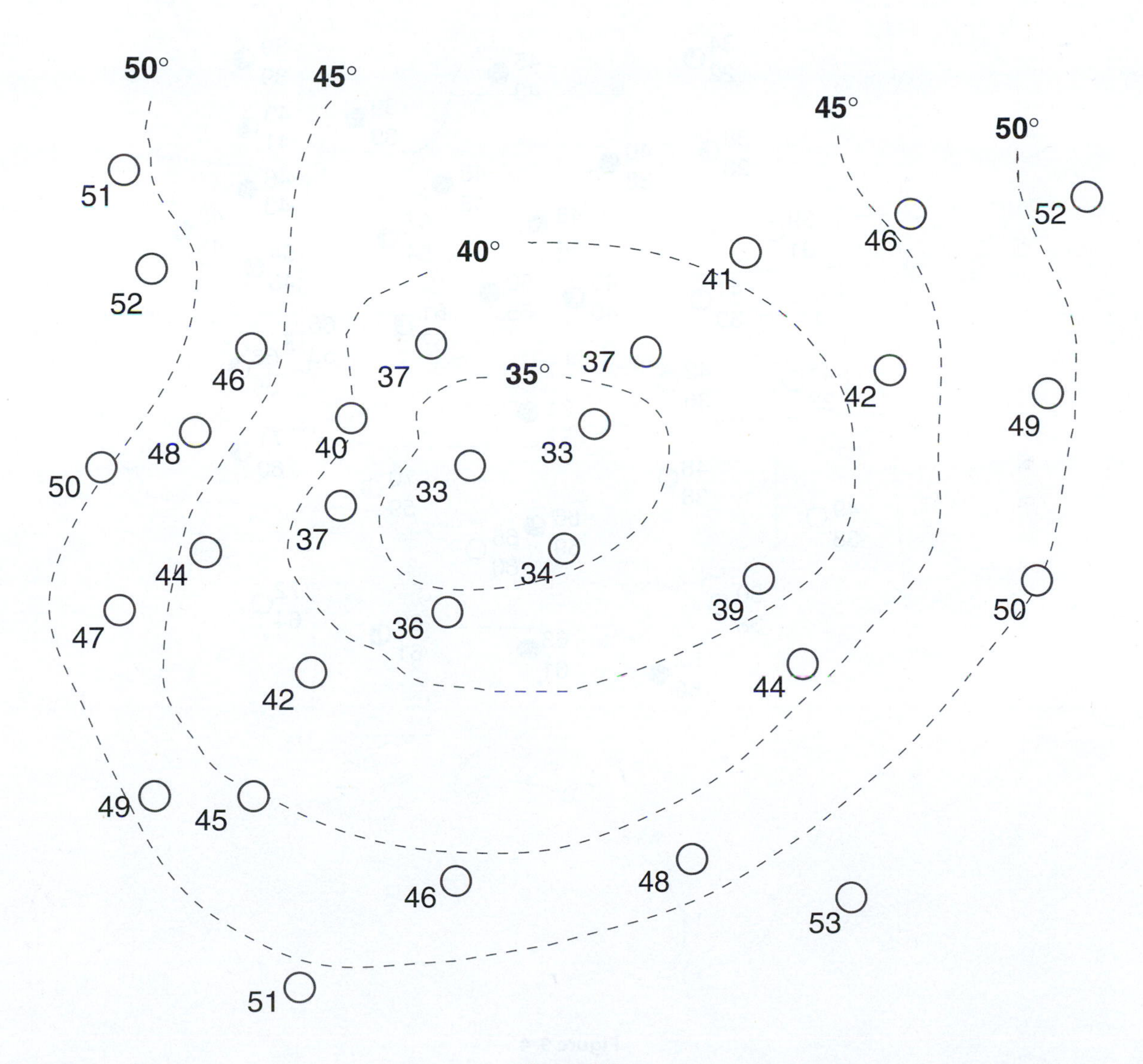

Figure 9-3

3. *Construct isotherms at 5°F intervals (e.g., 35°F, 40°F, 45°F) on the simplified weather map shown below. Use solid lines.*

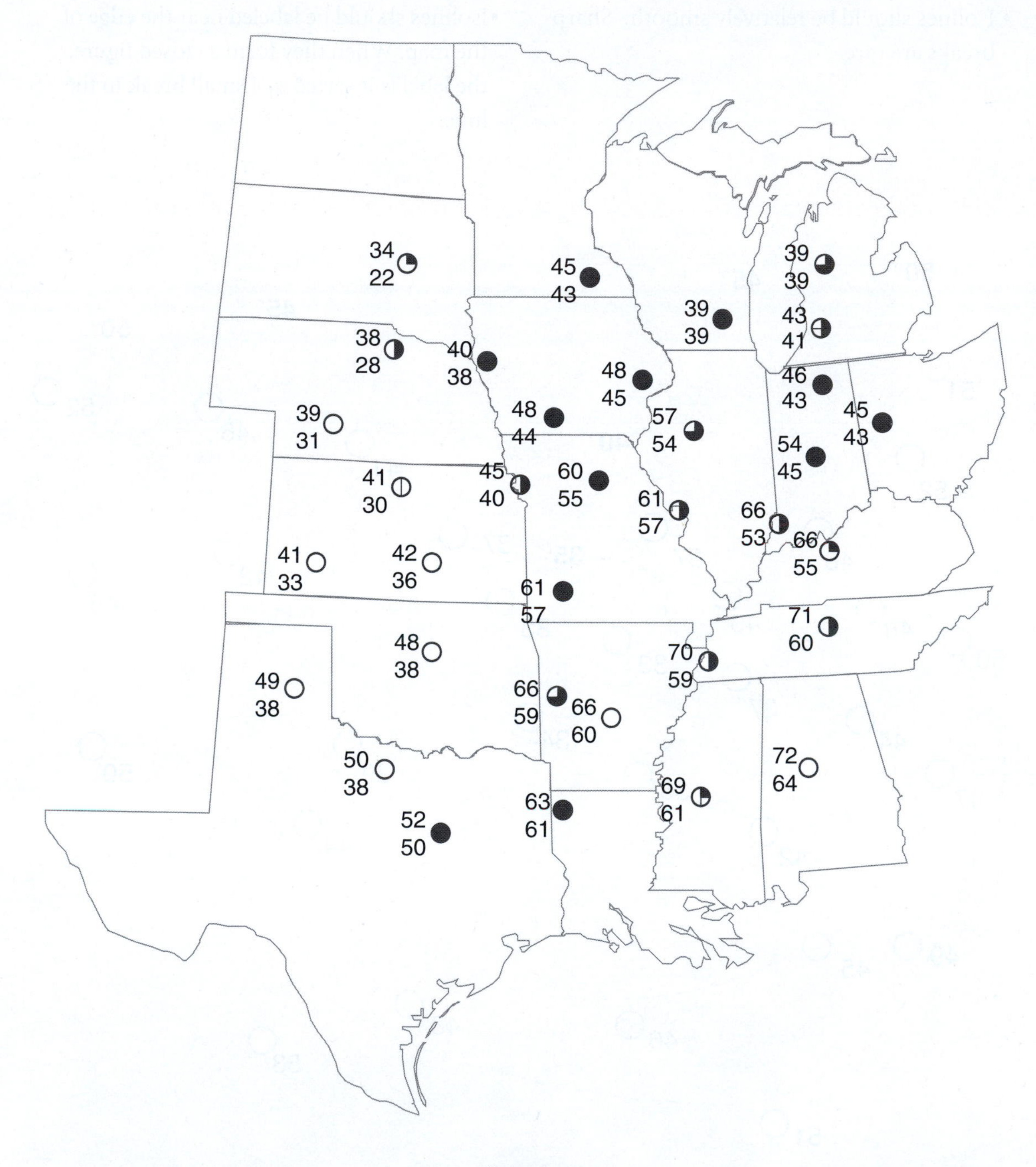

Figure 9-4

4. *Using dotted lines, now construct isodrosotherms at 5°F intervals on the map below.*

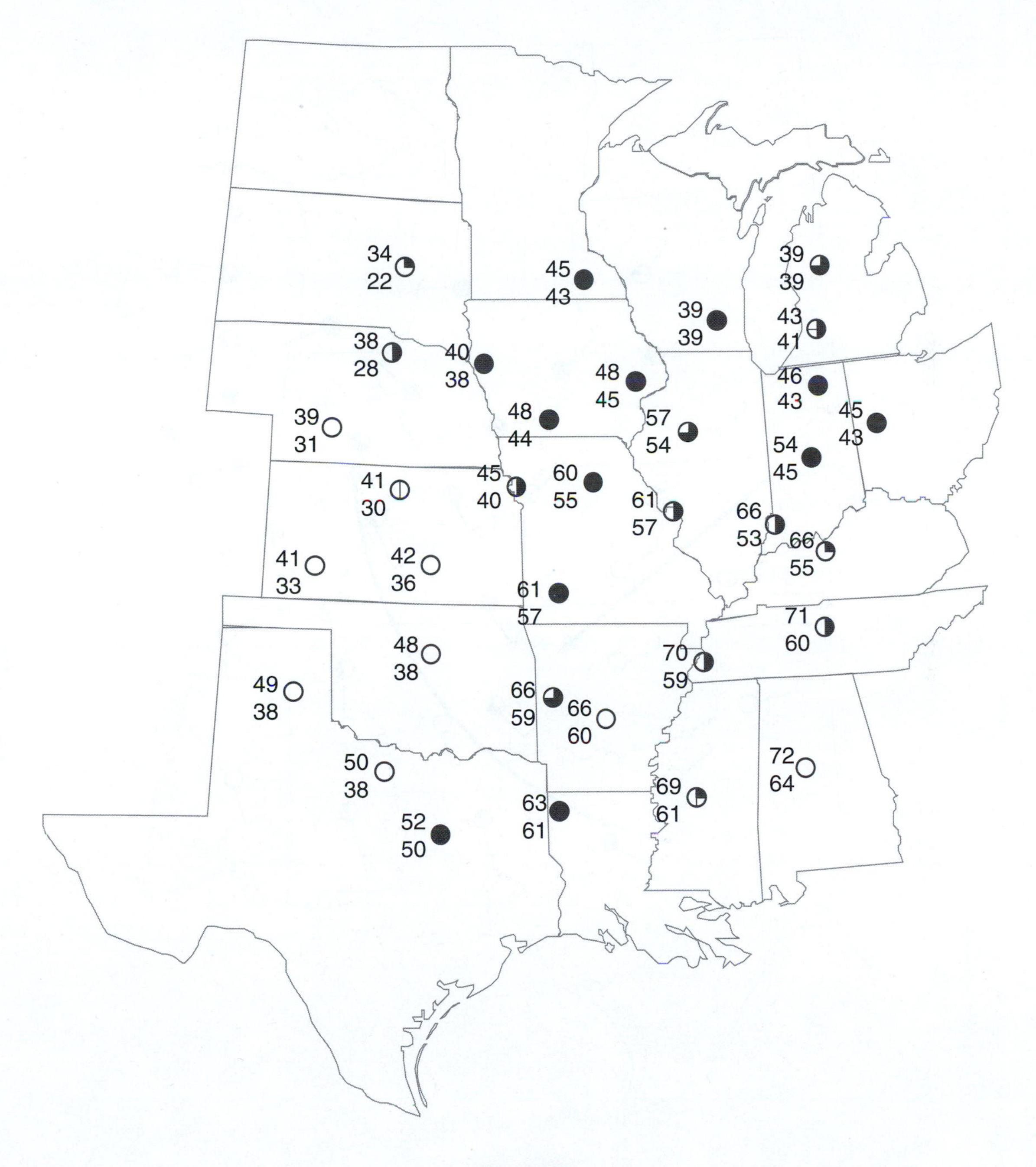

Figure 9-5

We can show wind flow patterns using *streamlines,* drawn parallel to the wind barbs used in the station model. Streamlines begin at an upwind location and are drawn as long lines ending with an arrow where the wind shifts abruptly (Figure 9-6).

5. *Complete the analysis below by drawing 5 to 6 streamlines to illustrate the general wind flow.*

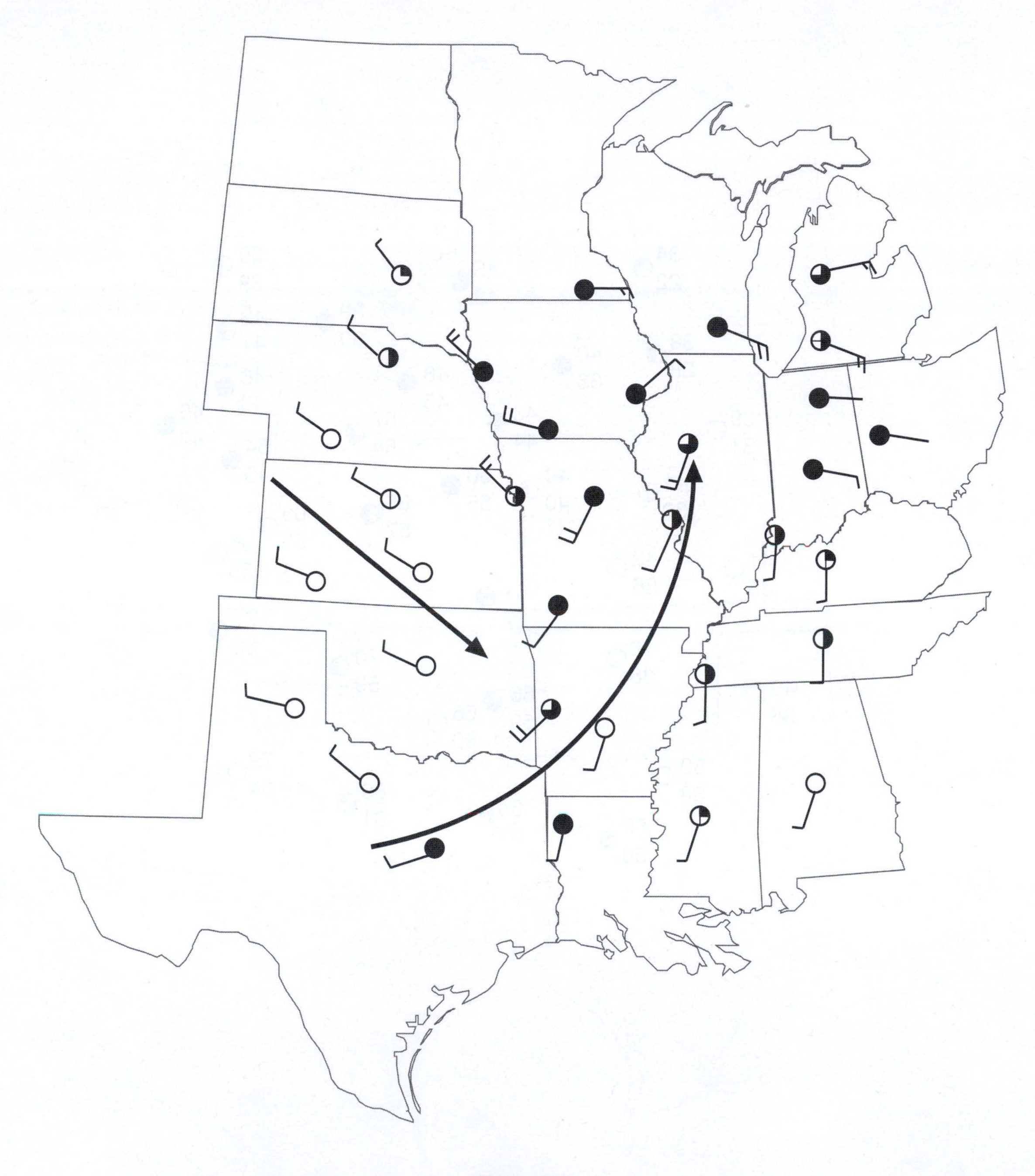

Figure 9-6

Air Masses and Fronts

An *air mass* is a large body of air with relatively uniform temperature and humidity characteristics. Air masses form over large land or water surfaces and take on the temperature and moisture characteristics of these surfaces, where they remain stationary for days or even weeks. Their moisture characteristics are classified as maritime or continental, and their temperature characteristics as equatorial, tropical, polar, or arctic. Maritime arctic and continental equatorial air masses are rarely found and therefore are not listed. Therefore, the following types of air masses result:

— maritime equatorial (mE)
— maritime tropical (mT)
— maritime polar (mP)
— continental tropical (cT)
— continental polar (cP)
— continental arctic (cA)

Air masses often migrate from their source regions and affect mid-latitude weather. Examine the diagram below showing air masses affecting North America.

6. *Based on the source regions shown by the ovals, indicate each type of air mass influencing North America (mT, mP, cT, cP, and cA).*

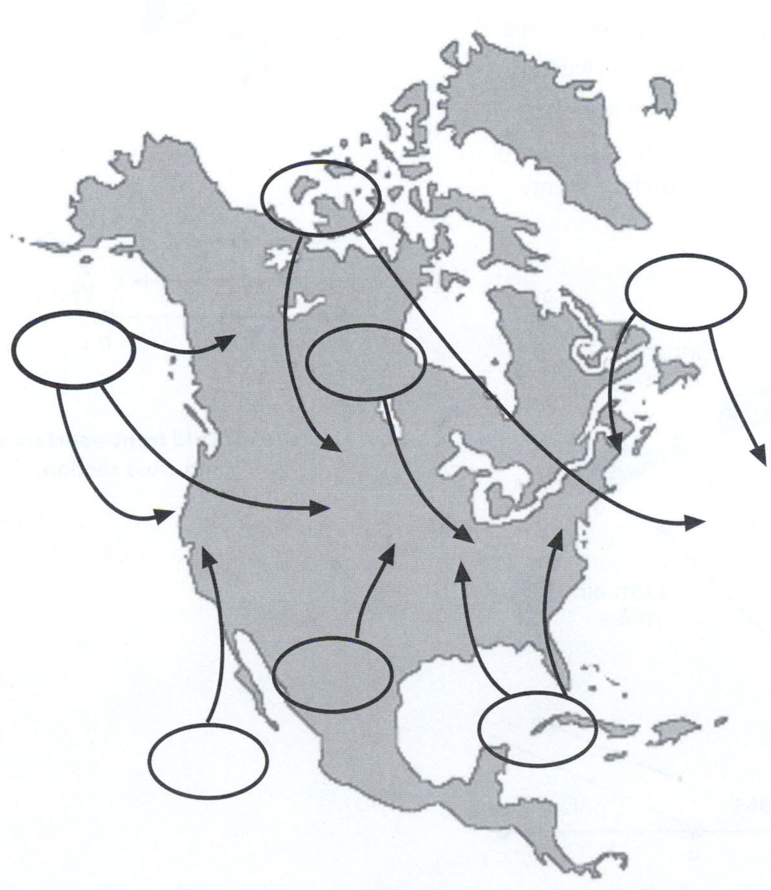

Figure 9-7

Fronts

A *front* marks the boundary between two unlike air masses. Fronts can be identified by any of the following characteristics: a sharp temperature gradient, a sharp moisture gradient, or a sharp change in wind direction. We categorize fronts according to their net movement. When air flows parallel to the boundary and neither air mass advances, the boundary is referred to as a *stationary front*.

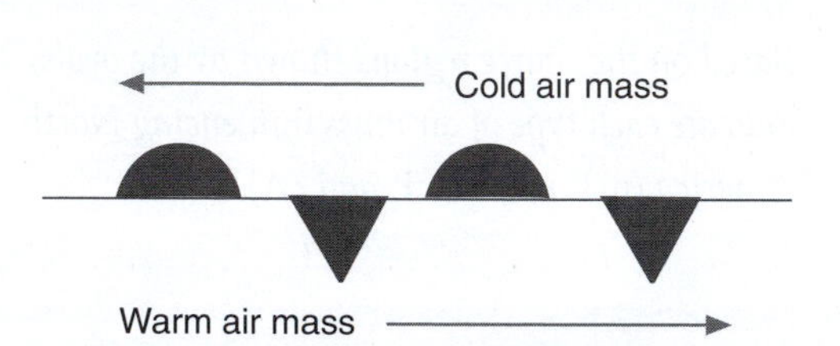

Figure 9-8. Stationary front—surface depiction.

When a warm air mass advances on a cooler air mass, the boundary between them is called a *warm front*. Because warm air is less dense, it will cool adiabatically and usually forms clouds ahead of the surface front.

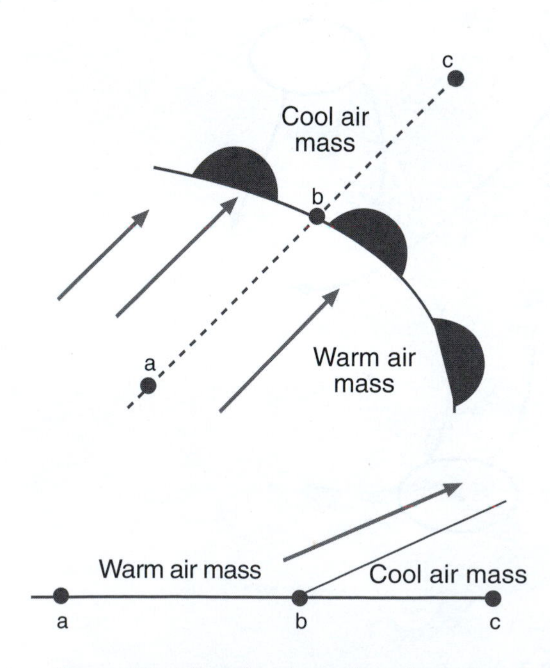

Figure 9-9. Warm front—surface depiction and cross section.

When a cold air mass advances on a warmer air mass, the boundary is called a *cold front*. In this case, cold air wedges itself beneath warm air because of its greater density. Surface friction creates a steep slope as the cold air advances. Since cold fronts generally move faster than warm fronts, warm air masses are lifted more rapidly along cold fronts and clouds grow to greater vertical extent than along most warm fronts.

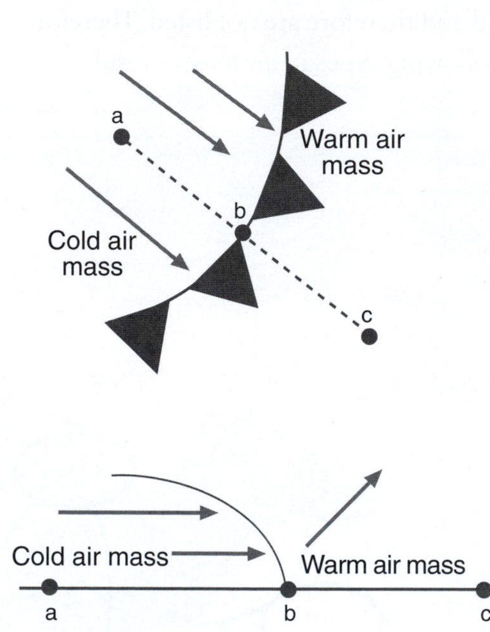

Figure 9-10. Cold front—surface depiction and cross section.

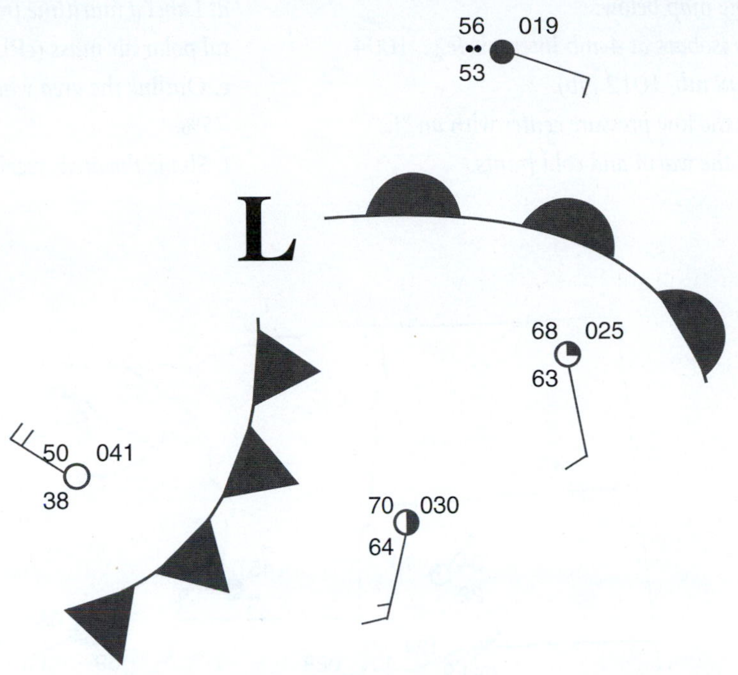

Figure 9-11

Usually, waves along a frontal boundary form a low pressure center. As air circulates around the low, warm and cold air masses advance, resulting in storm systems such as that illustrated in Figure 9-11. These fronts divide the contrasting air masses. Notice also how wind direction shifts across the frontal boundary.

As the wave amplifies, a cold front will often overtake a warm front. We define this new boundary as an *occluded front.*

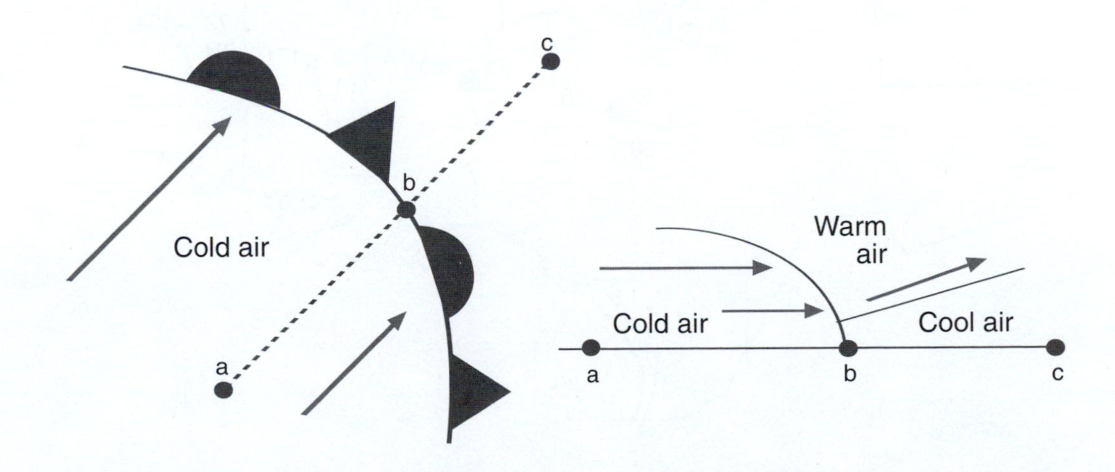

Figure 9-12. Occluded front—surface depiction and cross section.

7. *Using the map below:*
   a. Draw isobars at 4-mb intervals (e.g., 1004 mb, 1008 mb, 1012 mb).
   b. Label the low pressure center with an "L."
   c. Draw the warm and cold fronts.
   d. Label a maritime tropical (mT) and continental polar air mass (cP).
   e. Outline the area where cloud cover exceeds 75%.
   f. Shade the areas receiving precipitation.

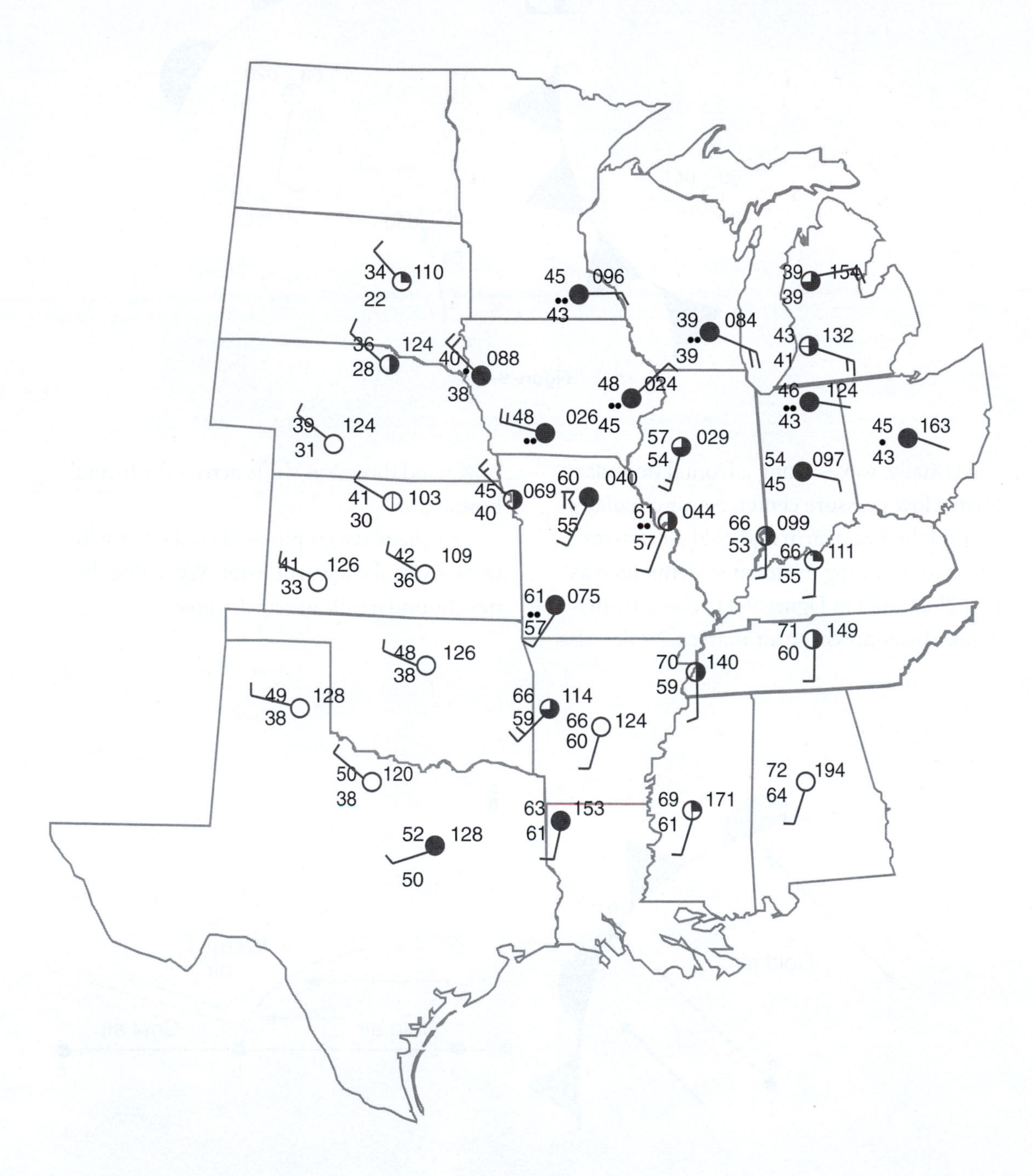

Figure 9-13

Upper-Air Charts

Upper-air charts depict spatial patterns of meteorological variables above the earth's surface. By convention, such maps are constructed at 0000z and 1200z (midnight and noon GMT) for specific pressure levels rather than fixed heights. To display pressure gradient force above the earth's surface, we examine the gradient of height, i.e., how the height of a given pressure level (e.g., 850-mb, 700-mb, 500-mb, 300-mb, 200-mb) changes over space. Recall from Lab 1 our rule-of-thumb: for every 5.6 km rise in height, pressure is reduced by half. By this convention, one would expect the 500-mb surface to be at approximately 5500–5600 m. This height will vary as a function of temperature (and, therefore, air density). The relationship between pressure and height allows us to use codes for height on upper-air maps. Figure 9-14 shows a sample data point from a 500-mb map and the conventions often used for height on all upper-air charts. These codes will help you interpret upper-air charts on the Internet.

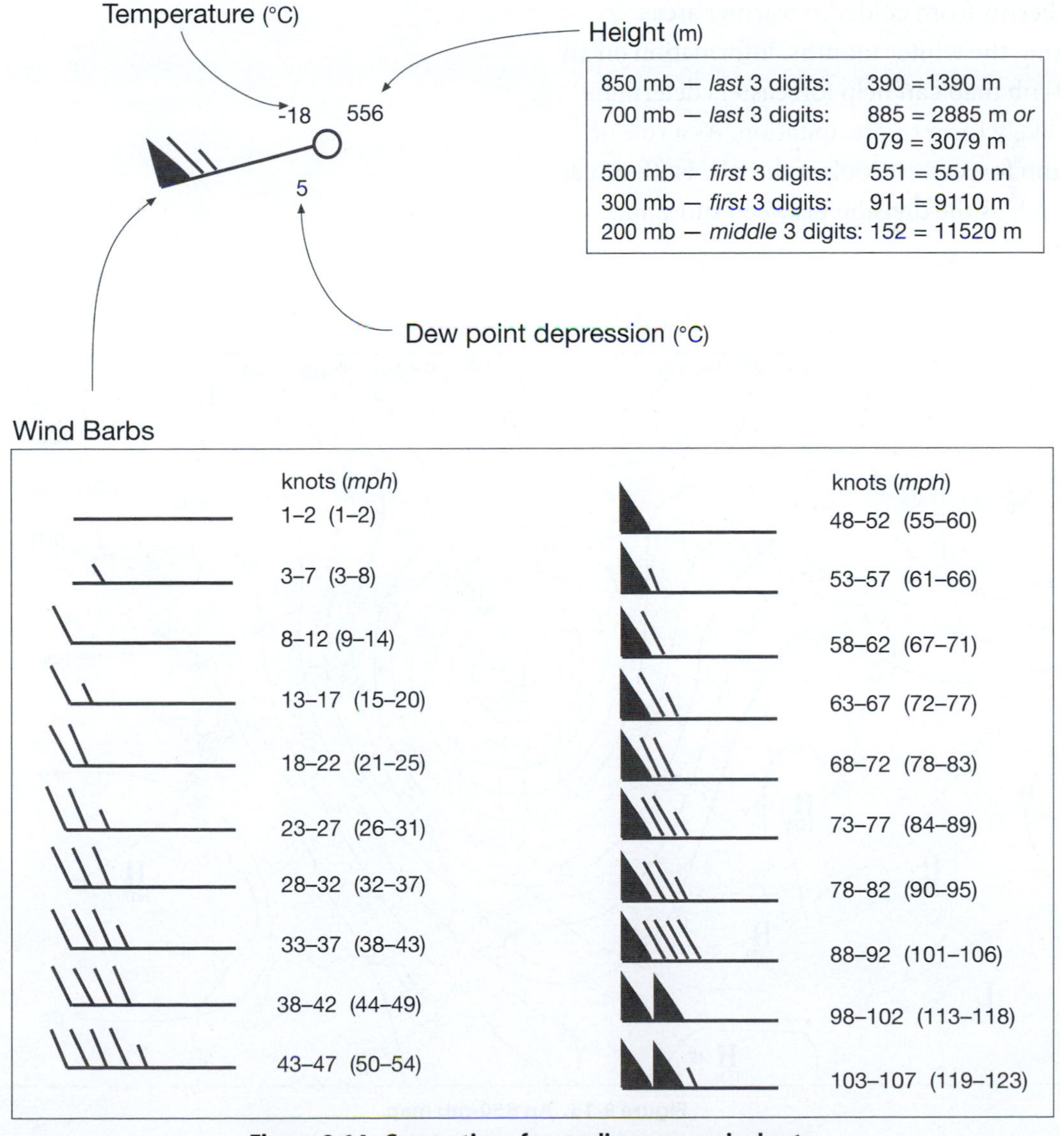

Figure 9-14. Conventions for reading upper-air charts.

850-mb maps

The 850-mb map provides a view of the atmosphere above the boundary layer where diurnal temperatures are strongly influenced by warming and cooling of the earth's surface. The 850-mb chart includes raw data at each station, height contours, and isotherms (Figure 9-15). Meteorologists use the 850-mb layer to find areas influenced by the movement of warm and cold air masses. Warm-air advection exists when winds blow across isotherms from warmer to colder areas; cold-air advection occurs where winds blow across isotherms from colder to warmer areas. During the winter months, information on an 850-mb map can help forecasters determine the likely form of precipitation. As a rule of thumb forecasters look at the 0 °C isotherm at 850 mb as the division between snow and rain.

8. *What range of heights do you see in this example of the 850-mb surface?*

9. *Circle and label areas of warm- and cold-air advection on the 850-mb map.*

10. *Highlight the 0 °C isotherm.*

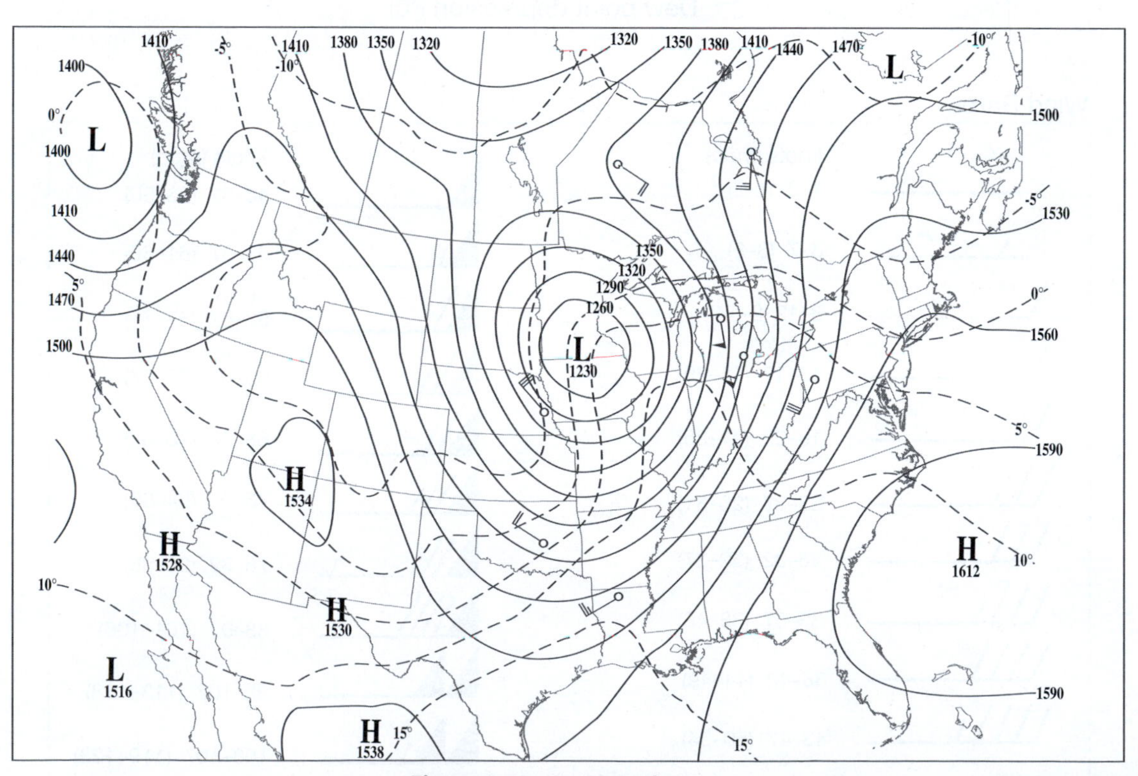

Figure 9-15. An 850-mb map.

500-mb maps

The 500-mb map generally includes the same raw data, contours, and isotherms as the 850-mb map. Occasionally, *vorticity* (a measure of eddies in the general wind flow) is also depicted. Maps at this level help to show how storms will move, since the speed and direction of 500-mb winds "steer" their paths.

11. *What contour interval is used to show heights on the 500-mb map?*

12. *What range of heights do you see on the 500-mb map?*

13. *How does the spatial pattern of heights illustrate the relationship between temperature, density, and the rate of vertical pressure change?*

14. *If a surface low pressure center developed in the Oklahoma panhandle, in which direction would it likely move?*

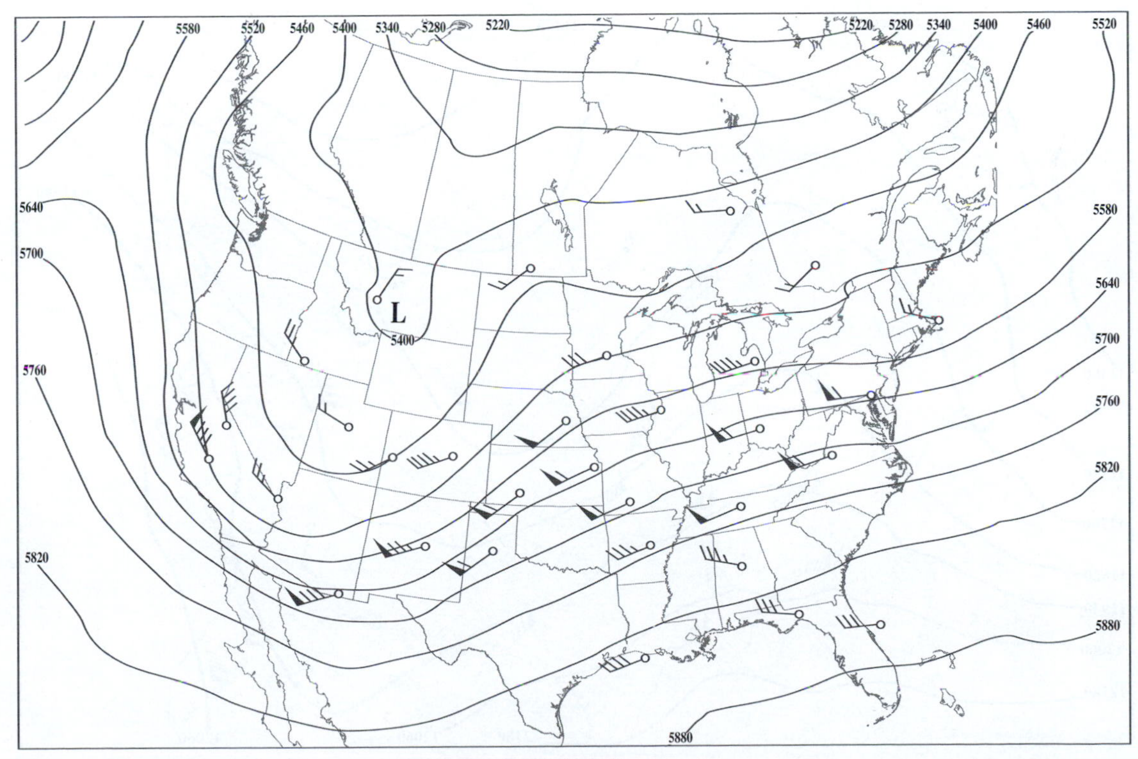

Figure 9-16. A 500-mb map.

300-mb and 200-mb maps

These maps help forecasters identify the jet stream. Typically, 300-mb maps are used during the coolest months, and 200-mb maps are used during warmest months because of seasonal variation in the jet-stream height. Sometimes 300- and 200-mb maps include shading to identify winds above 60 knots. In addition, the maximum jet stream winds, or jet streaks, are often identified with contours.

15. *What contour interval is used to show heights on the 200-mb map below?*

16. *Shade areas with winds exceeding 60 knots.*

17. *Circle the two regions in Figure 9-17 that have the fastest winds.*

Review Question

How does the drawing of isotherms, isodrosotherms, and streamlines help meteorologists identify potential areas of stormy weather?

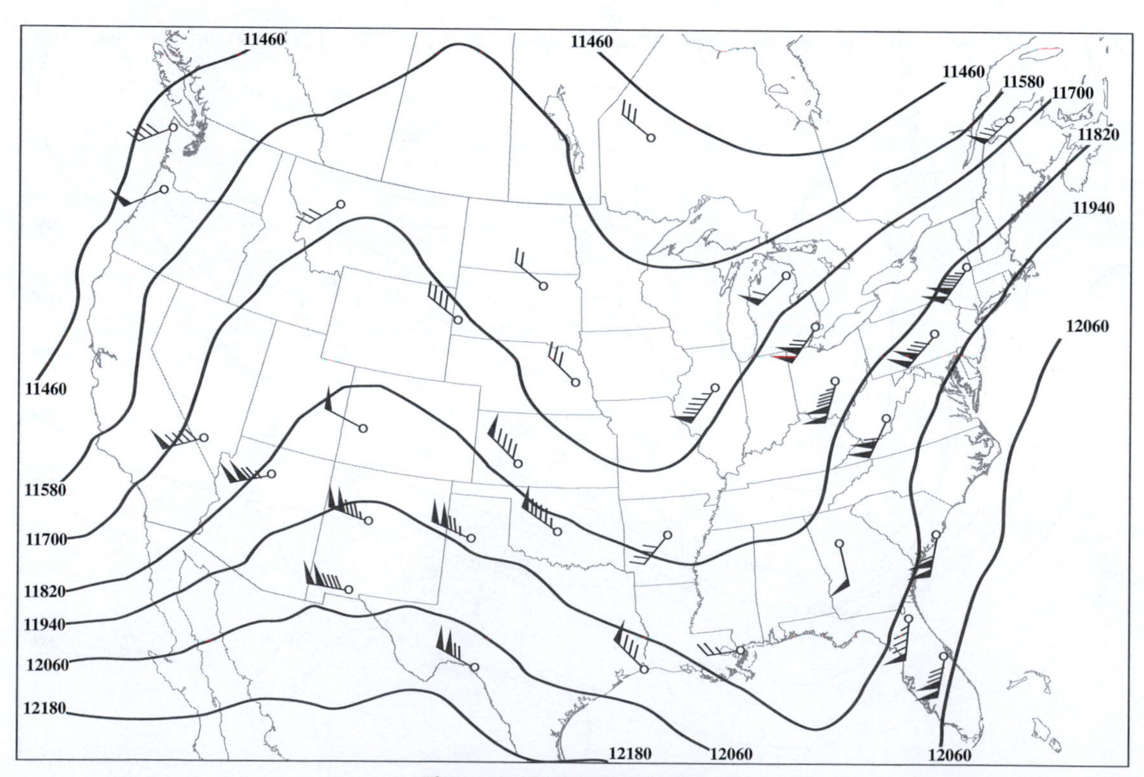

Figure 9-17. A 200-mb map.

Lab 10

Mid-Latitude Cyclones

Introduction

How do mid-latitude cyclones form and mature and how do these low pressure cells affect the weather? This lab uses idealized and real-world examples to illustrate the surface weather patterns associated with mid-latitude cyclones. It also examines how winds above the earth's surface influence storm processes, development, and movement.

The Polar Front Theory

We begin by examining the *Norwegian polar front theory,* which provides a model for mid-latitude cyclone development. In this model, cyclones form along a stationary front dividing a warm subtropical air mass from a cold polar air mass. The air masses have different densities and move in opposite directions on either side of the front, creating a wave (Figure 10-1a). Under some conditions this wave may amplify, allowing warm air to penetrate poleward and cold air to penetrate equatorward. This penetration can cause a low pressure center to develop along the wave (Figure 10-1b). In this stage warm air is forced upward along the cold and warm fronts and close to the low pressure center. Since cold fronts typically advance faster than warm fronts, the warm air sector is progressively lifted from the surface, and an occluded front forms (Figure 10-1c). This marks the most intense stage of the mid-latitude cyclone, characterized by maximum cloud cover and precipitation and the steepest pressure gradient. As air mass contrasts diminish, the storm loses its fuel supply and the storm dissipates.

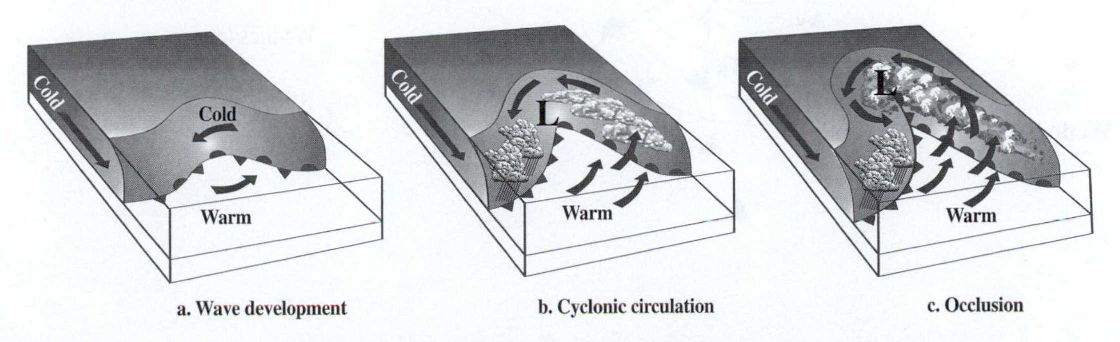

Figure 10-1. Bjerknes' Norwegian cyclone model.

Although the Norwegian model simplifies mid-latitude weather, it provides a starting point for understanding how these storm systems and their associated fronts affect the temperature, dew point, pressure, winds, clouds, and precipitation of the regions to which they move. Examine the following figures showing the development of a mid-latitude cyclone over a three-day period.

1. *Label air masses on each of the three maps.*

2. *On all three maps and at each station (A–D), use the standard wind symbol to depict wind direction.*

3. *The conditions at each station (A, B, C, and D) are shown below with station models for each day. Each column represents the same station over the three days. Match the appropriate station to each column.*

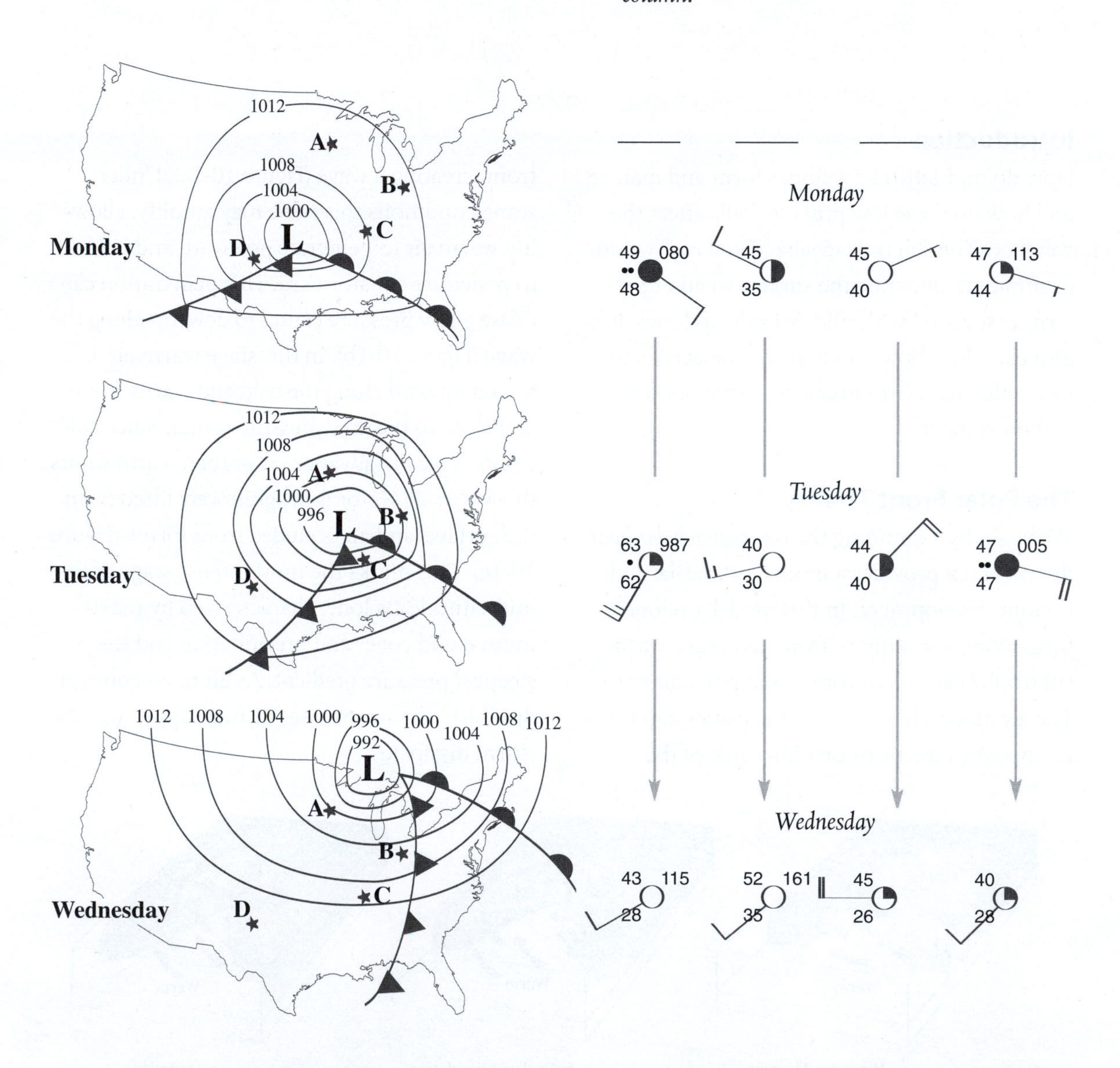

Figure 10-2

A Real-World Example

Few mid-latitude cyclones fit the classic Norwegian cyclone model precisely; however, intense mid-latitude cyclones provide useful case studies that link theories to the real world because their pronounced features reinforce key concepts related to storm development. Here we examine an example from November 1998 (Figures 10-3–10-6). This is a classic "Colorado low" forming in the lee of the Rockies. We will use upper-air and surface maps as well as time series of meteorological variables to investigate the storm's structure, how it developed, and how it affected specific places. Both Greenwich Mean Time (GMT) and Central Standard Time (CST) are given for each map.

4. *How much did the pressure drop in the storm's center from November 9, 1200z, until November 11, 0000z?*

5. *How did the surface pressure gradient over the upper Midwest change during this 36-hour period? How did surface wind speed respond to this pressure gradient change?*

Examine the cyclone on November 10, 1200z:

6. *How do the cloud and precipitation patterns match those of the Norwegian cyclone model?*

7. *What was the range of temperature and dew-point values within the maritime tropical air mass at that time?*

8. *What was the predominant wind direction in the maritime tropical air mass?*

9. *How much lower were temperatures and dew-point temperatures in the continental polar air mass west of the cold front?*

10. *Describe the differences in wind speed and direction between the eastern and western sides of the cold front.*

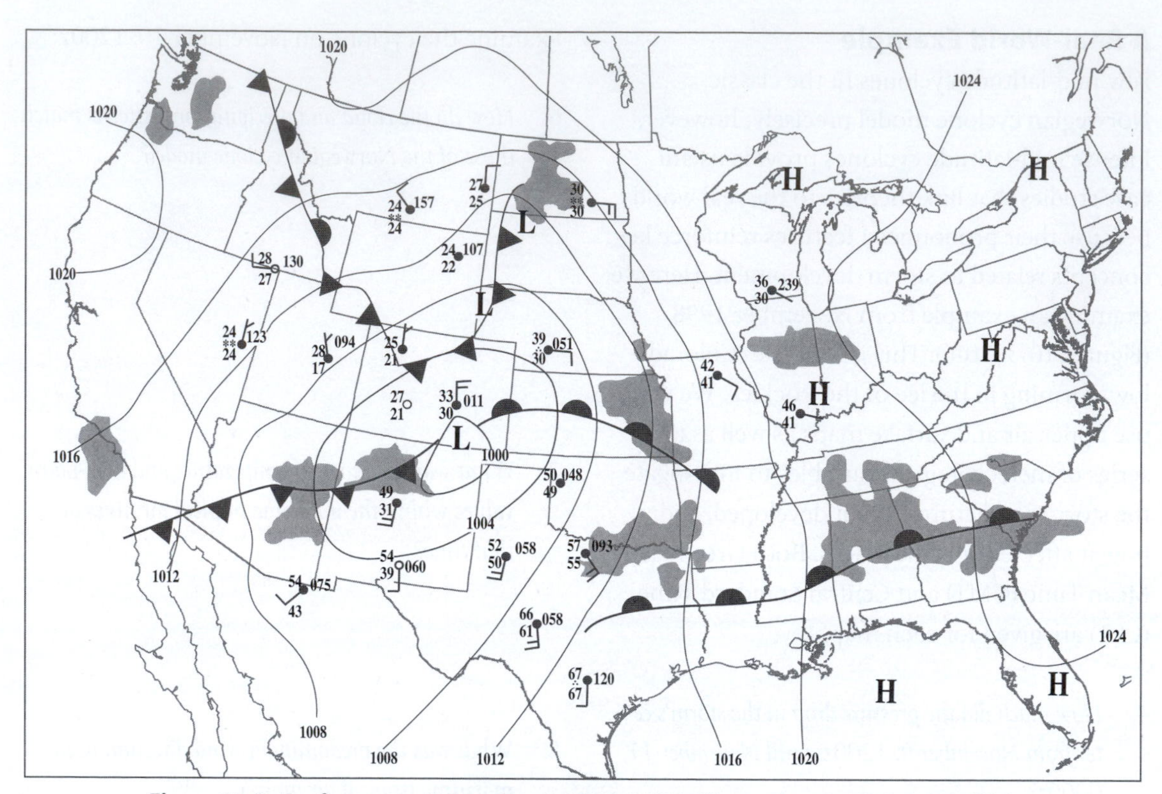

Figure 10-3. Surface map, November 9, 1998, 1200z (November 9, 1998, 0600 CST).

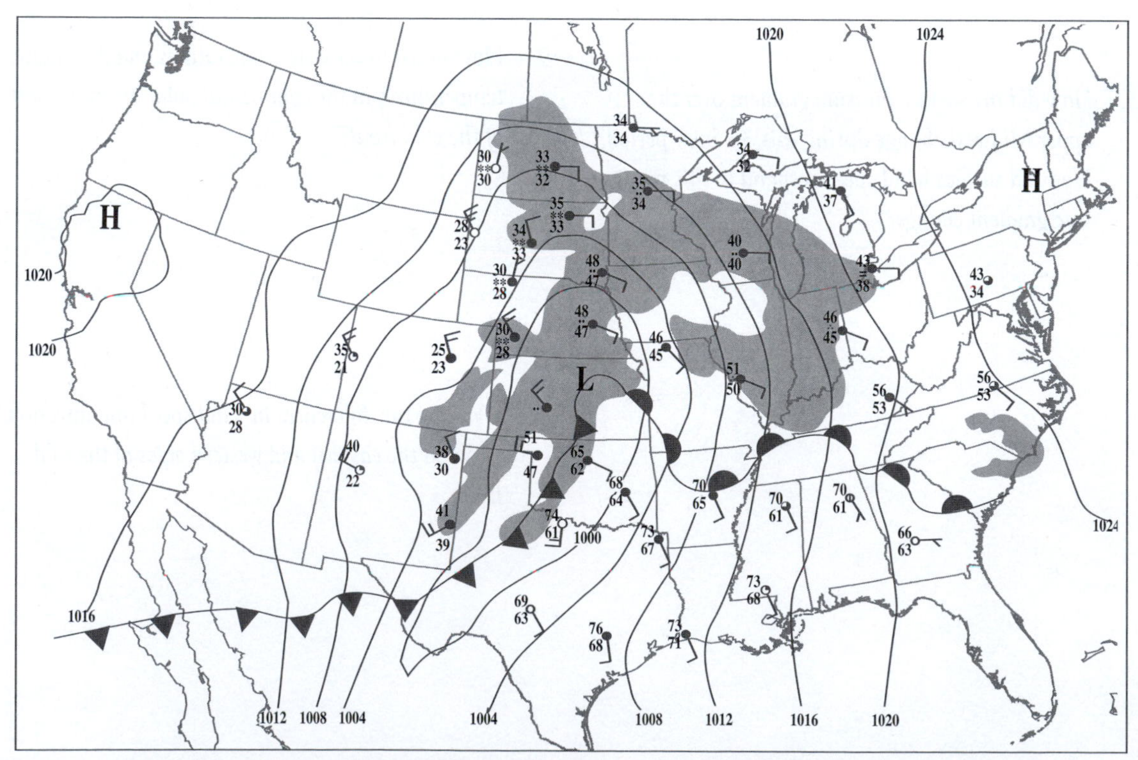

Figure 10-4. Surface map, November 10, 1998, 0000z (November 9, 1998, 1800 CST).

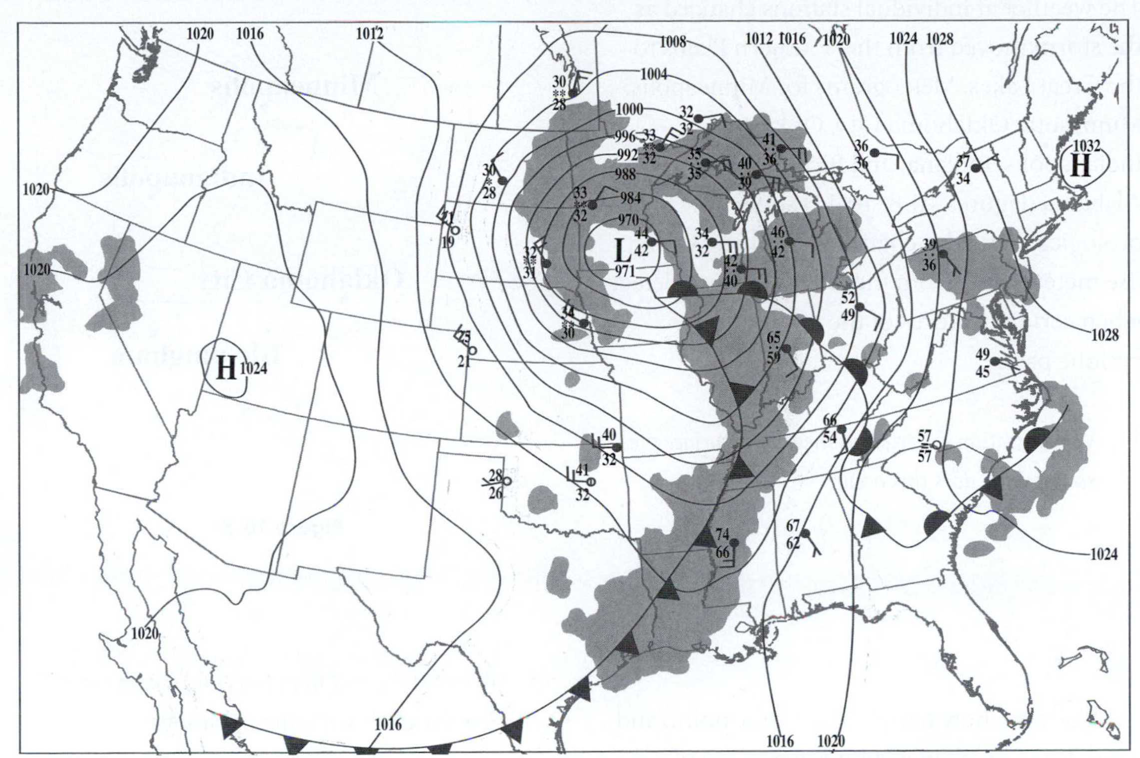

Figure 10-5. Surface map, November 10, 1998, 1200z (November 10, 1998, 0600 CST).

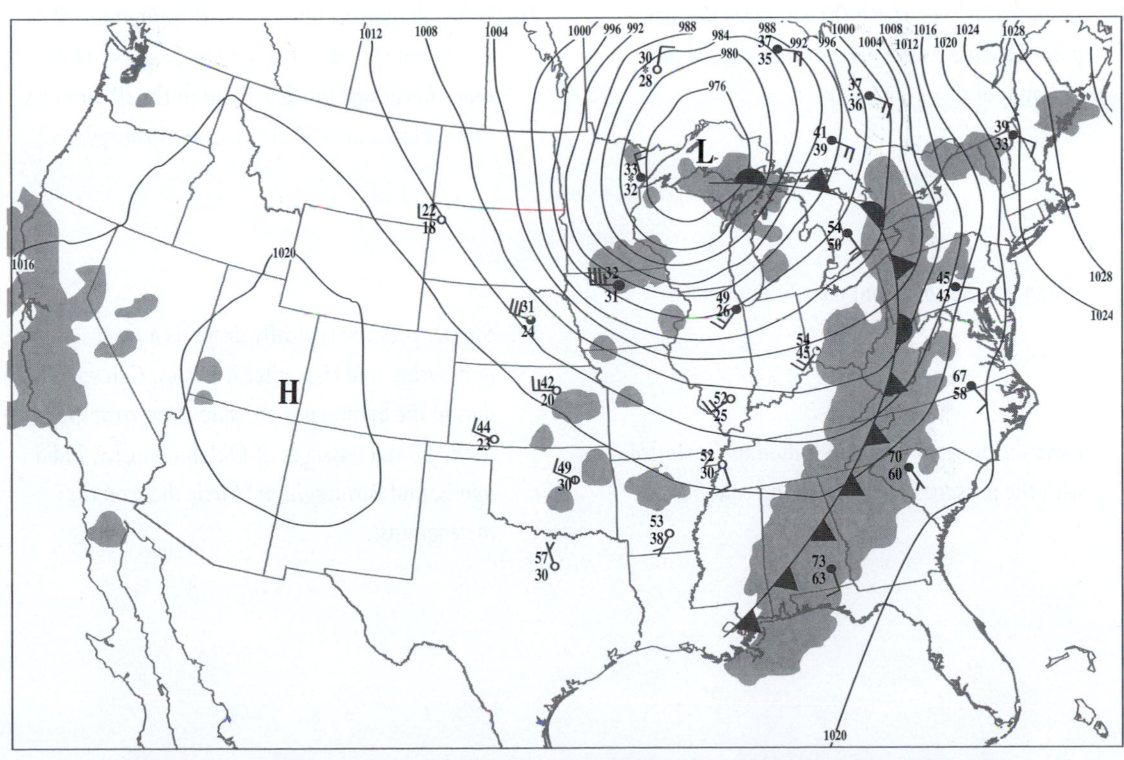

Figure 10-6. Surface map, November 11, 1998, 0000z (November 10, 1998, 1800 CST).

The weather at individual stations changed as the storm moved from the Southern Plains to the Great Lakes. Meteograms for Minneapolis, Minnesota; Oklahoma City, Oklahoma; Indianapolis, Indiana; and Birmingham, Alabama (Figures 10-8–10-11) show how meteorological variables change with time. We can use meteograms from these four cities to detect when certain features of the mid-latitude cyclone pass.

Figure 10-7

11. *Which station experiences the lowest surface pressure? When does this occur?*

Consider how temperature, dew point, and wind direction change with the passage of a front and reconcile your answers with respect to the surface weather maps.

12. *From 9/0300 to 10/0900 what was the dew-point increase as the mT air mass moved into Indianapolis?*

13. *When did the cold front pass Indianapolis?*

14 *Describe how the winds in Indianapolis shifted with the passage of the warm and cold fronts.*

15. *When did the cold front pass Oklahoma City? How did winds shift after its passage?*

16. *When did the cold front pass Birmingham? Was there rainfall associated with its passage? How much lower was the dew point in the cP air mass that replaced the mT air mass in Birmingham?*

17. *Surface pressure typically drops as a front approaches and rises after it passes. Can you detect dips in the barometric pressure curve corresponding to the frontal passages at Oklahoma City, Indianapolis, and Birmingham? Circle these on the meteograms.*

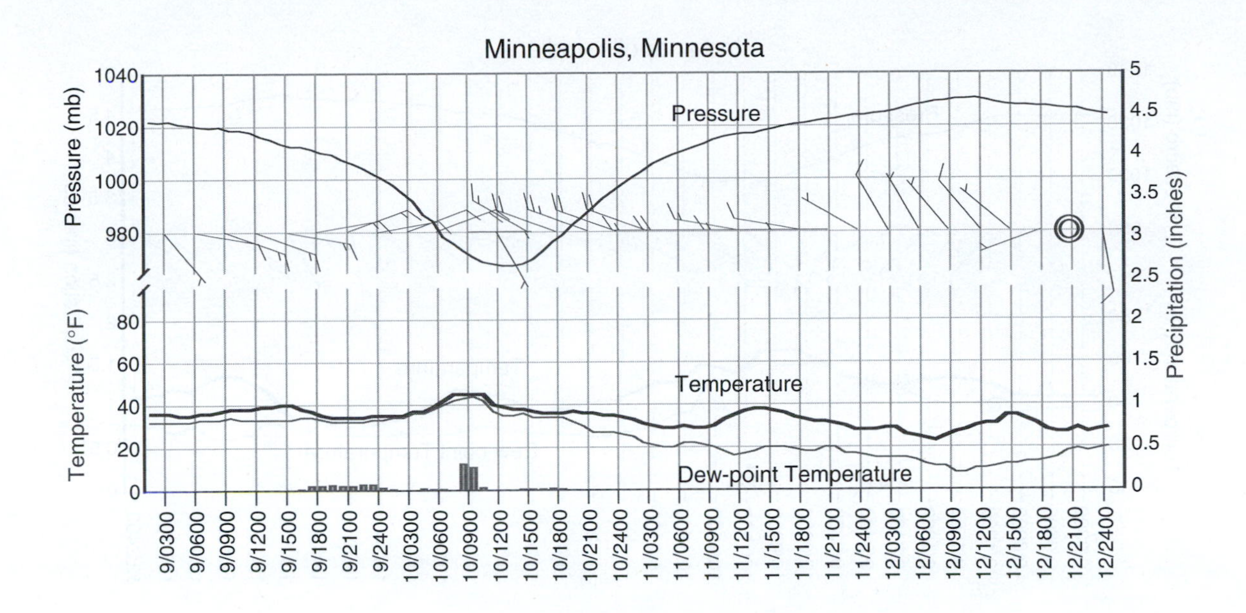

Figure 10-8. Meteogram for Minneapolis, Minnesota, November 9–12, 1998. Times are Central Standard Times (6 hours earlier than GMT).

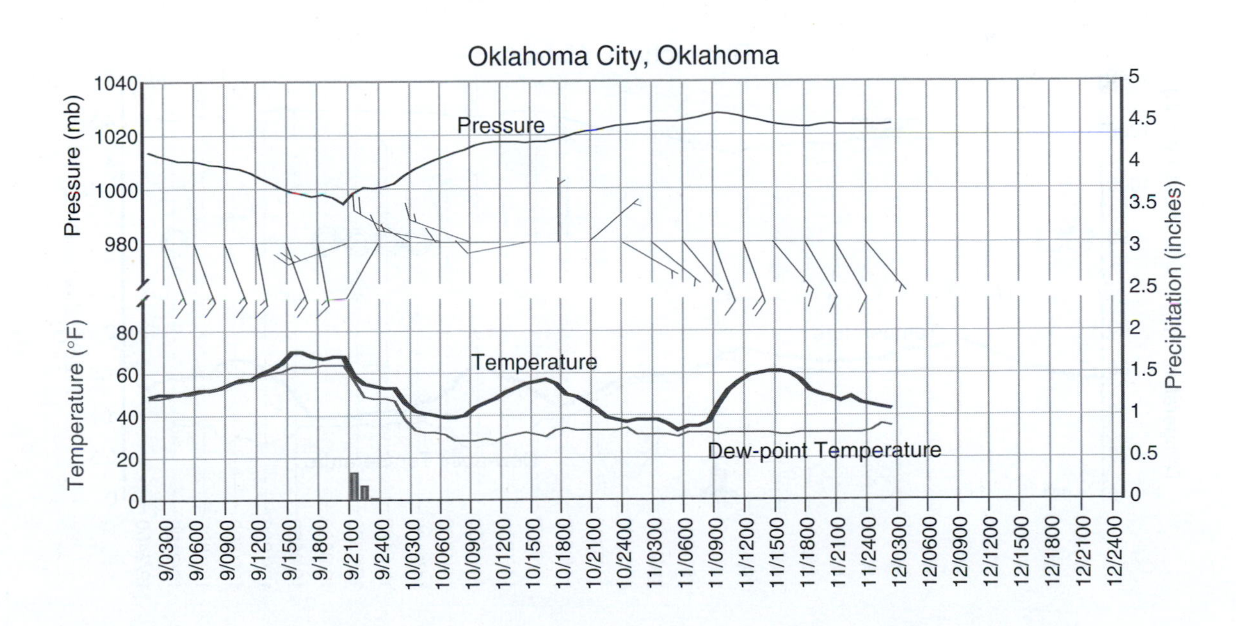

Figure 10-9. Meteogram for Oklahoma City, Oklahoma, November 9–12, 1998. Times are Central Standard Times (6 hours earlier than GMT).

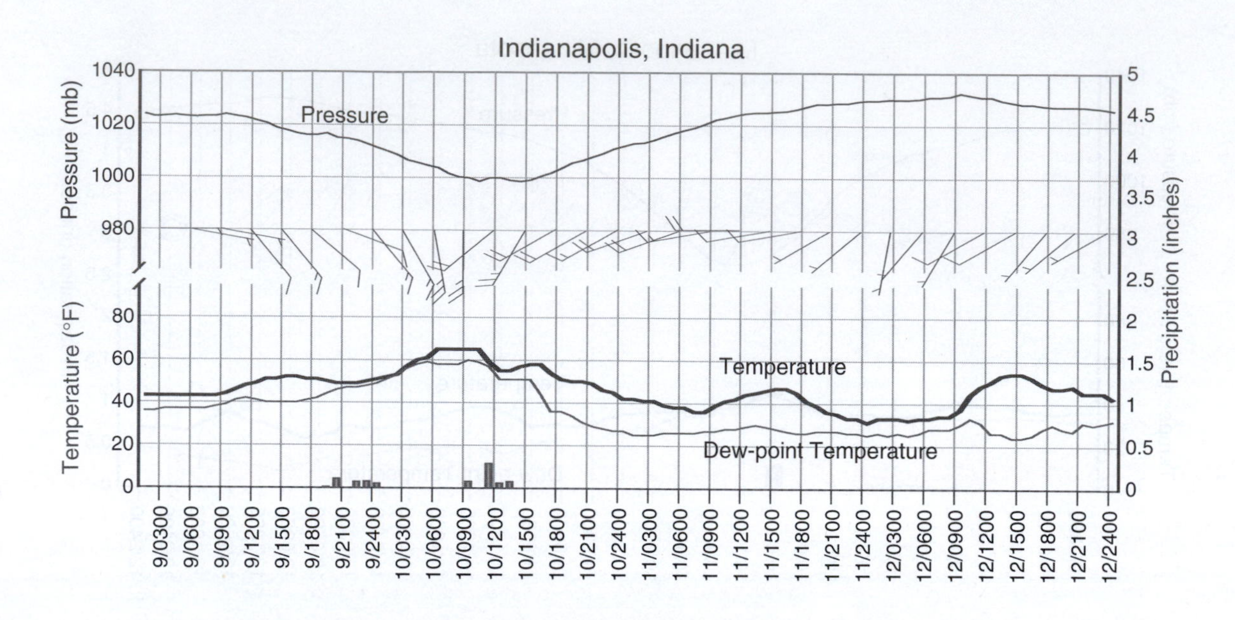

Figure 10-10. Meteogram for Indianapolis, Indiana, November 9–12, 1998. Times are Central Standard Times (6 hours earlier than GMT).

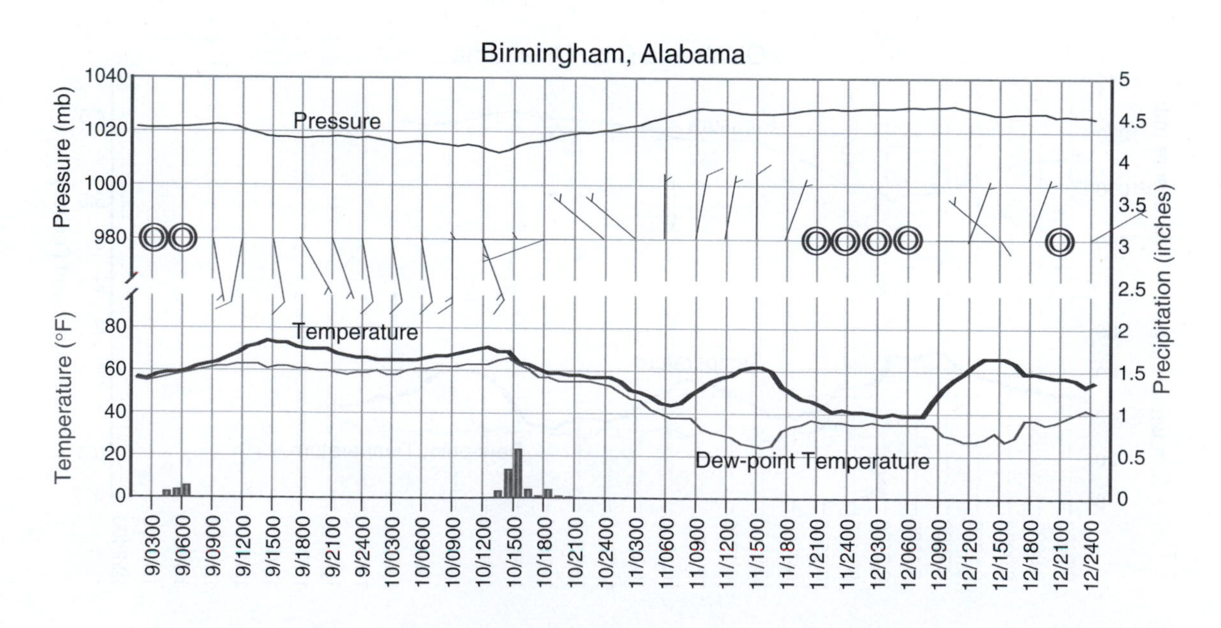

Figure 10-11. Meteogram for Birmingham, Alabama, November 9–12, 1998. Times are Central Standard Times (6 hours earlier than GMT).

Upper-Air Analysis

Of course, we must consider air flow above the earth's surface to understand why mid-latitude cyclones produce precipitation. Upper-air maps, produced every 12 hours, help us understand the relationship between surface weather and upper-level circulation.

850-mb

Examine the November 10, 1200z, 850-mb map showing wind flow and temperature patterns during the storm at approximately 1.5 km above sea level. The solid lines are contours, indicating constant height. The dashed lines are isotherms —lines of constant temperature that help us identify frontal boundaries above the earth's surface. When mapped together, contours and isotherms can help us identify areas of warm-air advection and cold-air advection. If we assume that winds flow parallel to contour lines, then we can find such areas where the winds blow across isotherms.

18. *Circle and label areas of warm-air and cold-air advection on the November 10, 1200z, 850-mb map.*

19. *Air should rise in areas of warm-air advection and sink in areas of cold-air advection. Based on these criteria, describe the direction (up or down) and location of vertical motion with respect to the surface warm and cold fronts.*

20. *How does the area of warm-air advection and upward vertical motion help explain the precipitation pattern found on the November 10, 1200z, surface map?*

21. *Highlight the 0° isotherm. Does it conform to the rain-snow boundary seen on the surface map for November 10, 1200z?*

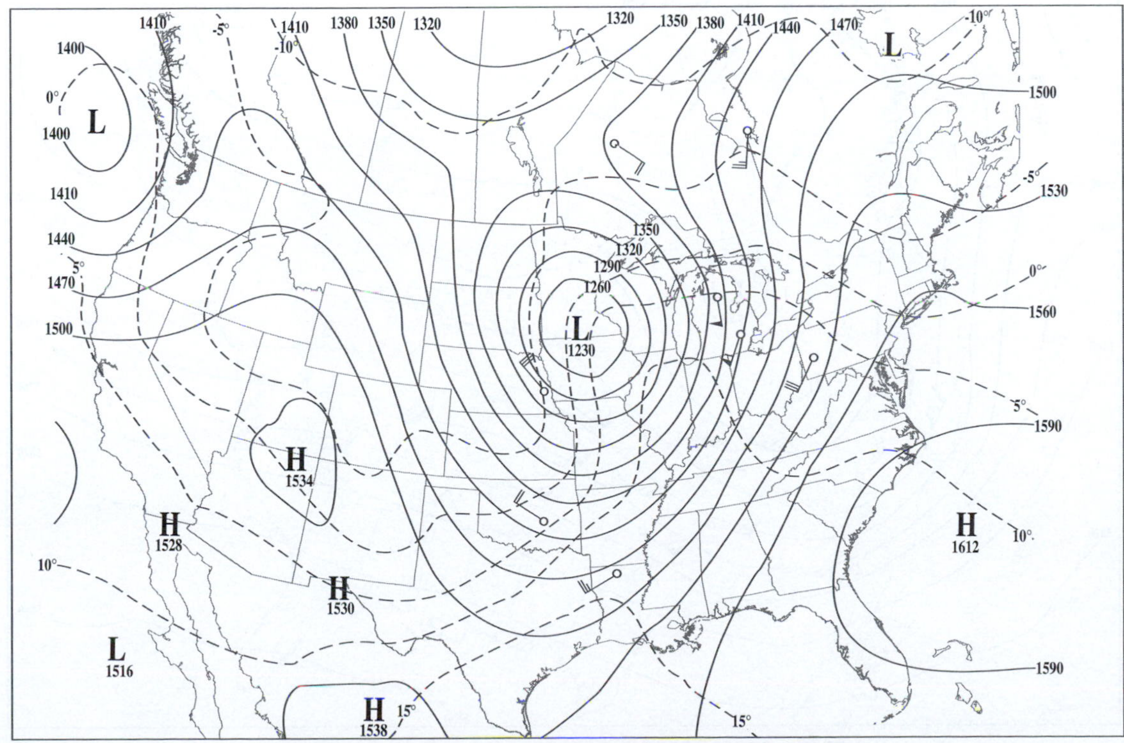

Figure 10-12. 850-mb map, November 10, 1998, 1200z.

500-mb

The 500-mb winds are said to "steer" mid-latitude cyclones, because they cause upper-air divergence that maintains surface low pressure. Areas of directional divergence commonly help forecasters predict future storm movement.

22. *Identify an area of directional divergence (diffluence) on the 500-mb map that contributed to the storm's development and path.*

23. *The surface low moved from southeastern Colorado to north-central Kansas (375 miles) from November 9th, 1200z, to November 10th, 0000z. How did this speed compare with the 500-mb wind speeds in this region?*

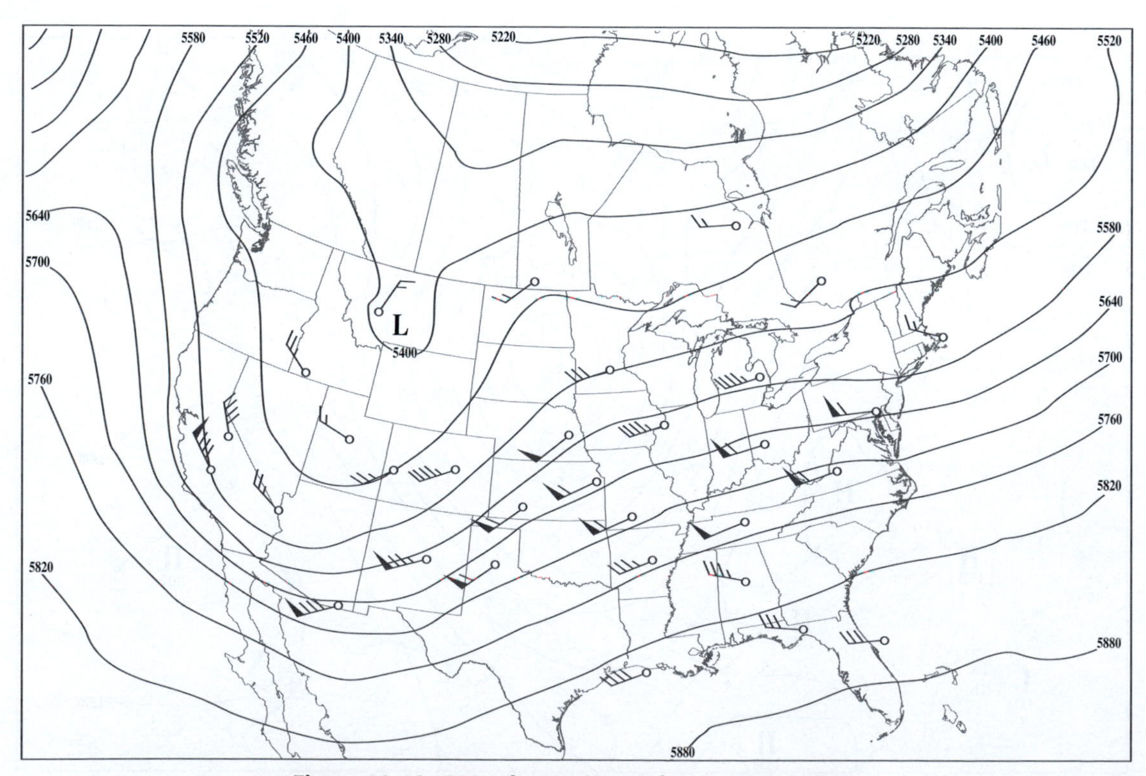

Figure 10-13. 500-mb map, November 9, 1998, 1200z.

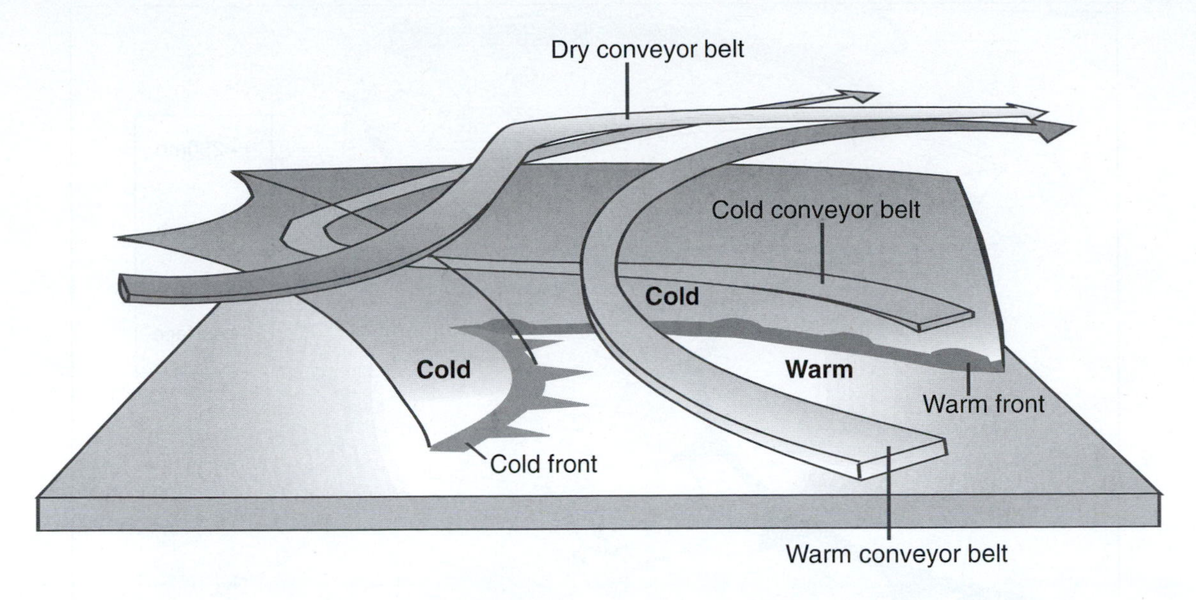

Figure 10-14. The conveyor-belt model.

The Conveyor-Belt Model

The wind flow associated with mid-latitude cyclones can be characterized by three major flows, referred to as *conveyor belts* (Figure 10-14). The warm conveyor belt originates as a surface wind in the warm-air sector of the mid-latitude cyclone. Because of its buoyancy, it rises over cooler air in the middle troposphere and eventually becomes westerly in the upper troposphere. The cold conveyor belt also begins near the surface as an easterly wind. As it rises it veers clockwise and also, ultimately, becomes westerly. The dry conveyor belt relates most closely to jet-stream winds.

How well does the November 1998 storm fit the conveyor-belt model? By using selected upper-air stations, we can reconstruct the three-dimensional flow associated with the storm. The arrows drawn in Figure 10-15 represent wind direction taken from 250-mb, 500-mb, and surface maps on November 10, 1200z (Figures 10-16, 10-17, and 10-18, respectively). Collectively, these arrows show the cold conveyor belt.

24. *Following the example for the cold conveyor belt, now draw arrows depicting wind direction for the open ovals in Figure 10-15 to show the warm and dry conveyor belts.*

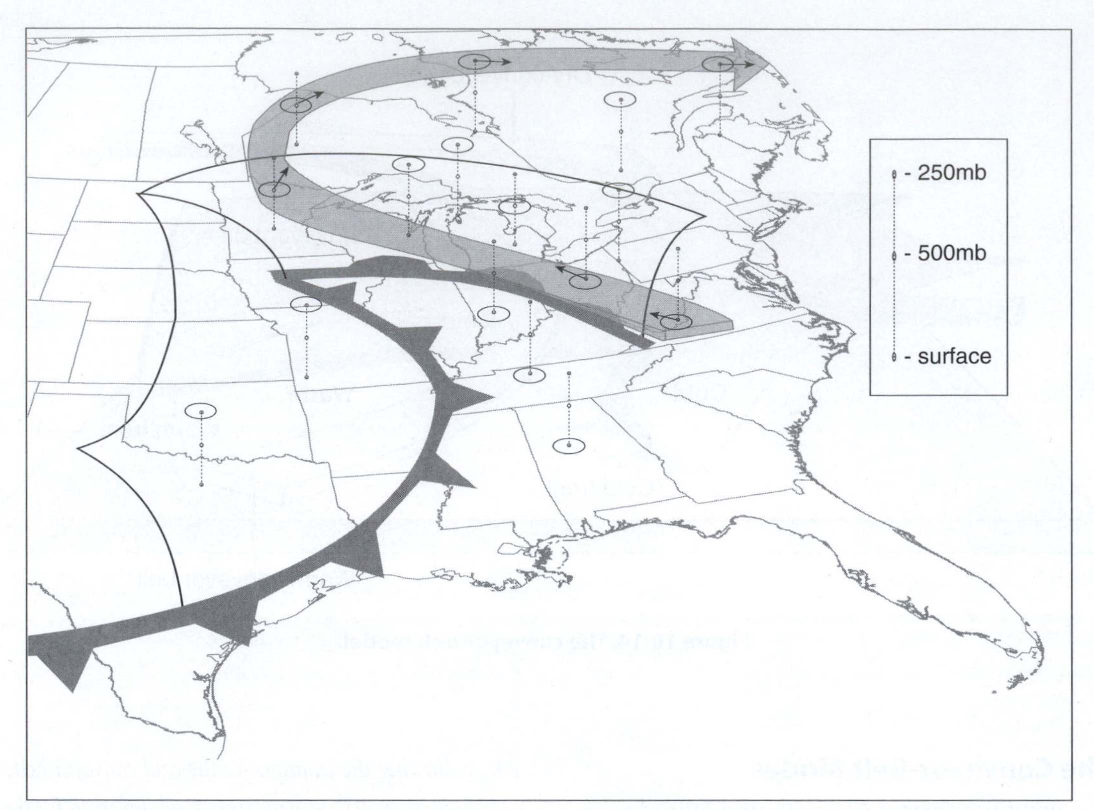

Figure 10-15. The conveyor-belt model applied on November 10, 1998, 1200z.

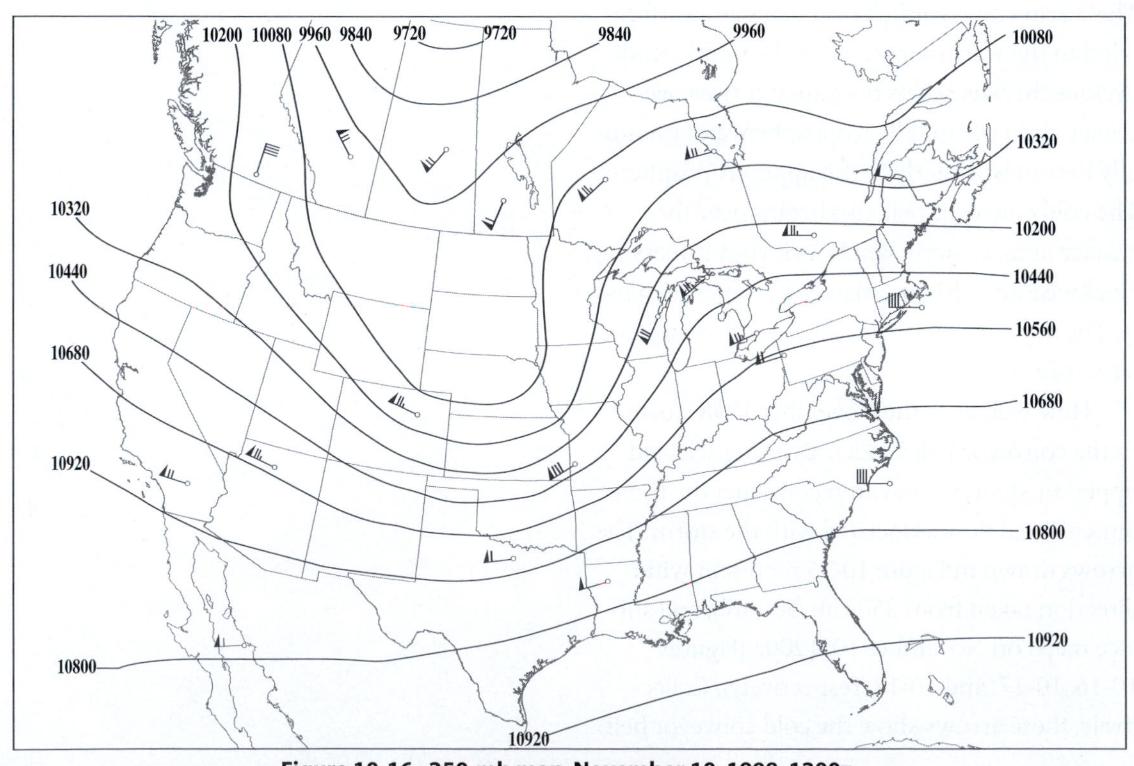

Figure 10-16. 250-mb map, November 10, 1998, 1200z.
Contour lines labeled in meters.

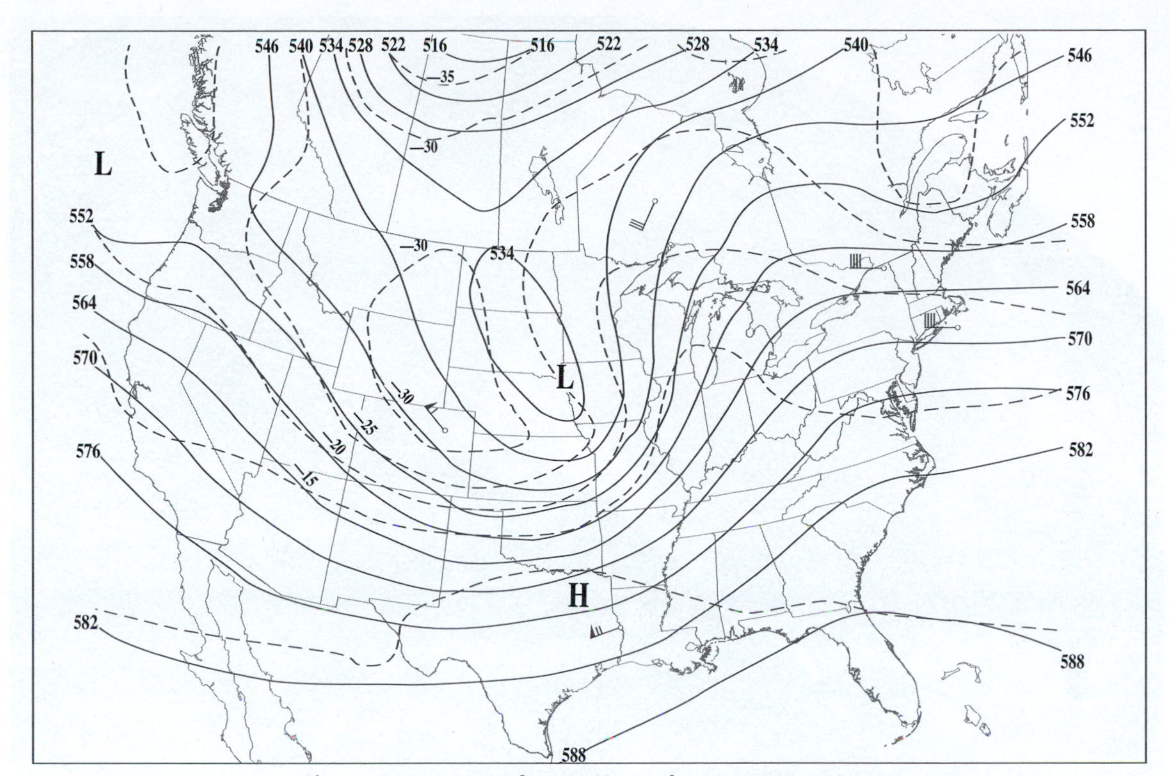

Figure 10-17. 500-mb map, November 10, 1998, 1200z.

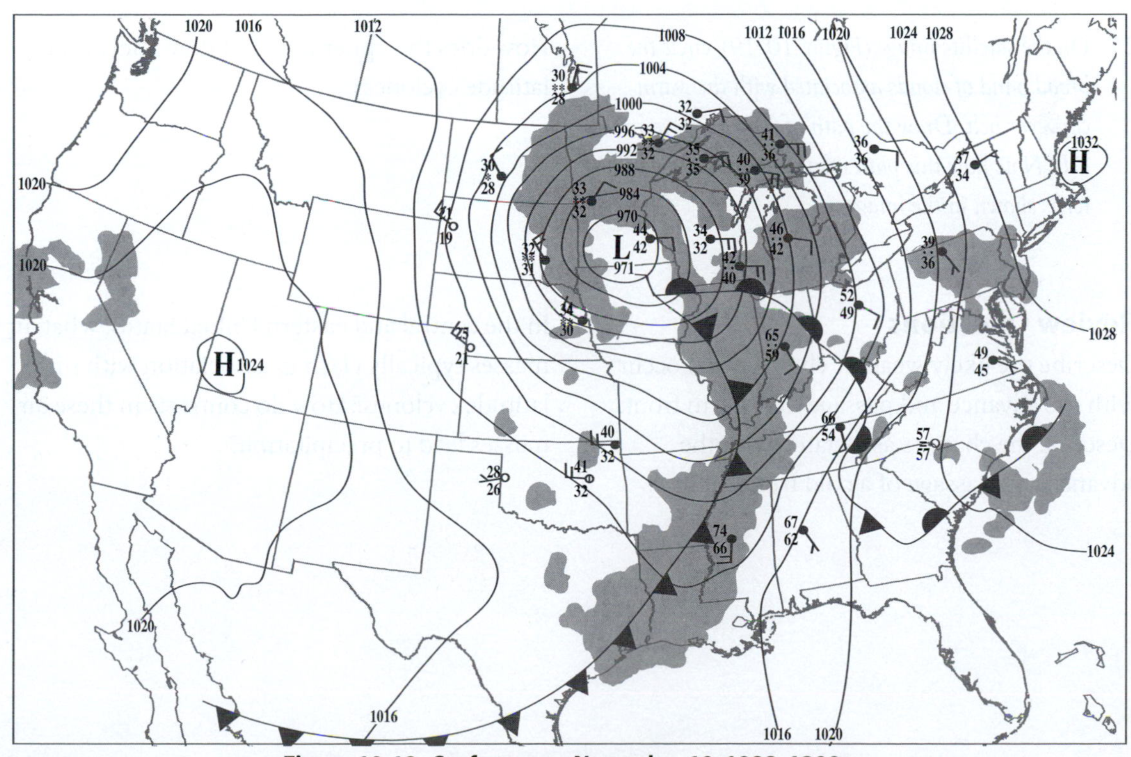

Figure 10-18. Surface map, November 10, 1998, 1200z.

Figure 10-19. Infrared satellite image, November 10, 1998, 1215z. Source: National Oceanic and Atmospheric Administration

25. *On the satellite image (Figure 10-19), circle the broad band of clouds associated with the warm conveyor belt. Draw the path of the dry conveyor belt. Note how this path relates to the cloud patterns shown in the image.*

Review Questions

Describe the likely weather changes that occur with the advance and passage of a warm front. Describe the changes associated with the advance and passage of a cold front.

How does the upper-air wind flow affect mid-latitude cyclones?

In the central and eastern United States, what air masses typically clash in association with mid-latitude cyclones? How do contrasts in these air masses lead to precipitation?

Lab 11

Thunderstorms and Tornadoes

Introduction

What causes thunderstorms and tornadoes? These dramatic events develop when warm, moist air is forced to rise in an unstable atmosphere. Thunderstorms are often categorized as *air mass thunderstorms* or *severe thunderstorms*. The distinction recognizes that some develop within a warm, moist air mass, whereas others require strong vertical wind shear. We will focus on the geographic patterns of these storms, their structure, and the atmospheric conditions that produce them.

Spatial Patterns of Thunderstorms

In the United States, thunderstorms most frequently occur when maritime tropical air is forced aloft through orographic lifting, frontal lifting, intense surface heating, or convergence of surface air flow. Figure 11-1 shows the pattern of thunderstorm frequency in the United States.

1. *What is the major source of maritime tropical air affecting the United States? How does proximity to this source influence the spatial pattern of thunderstorms?*

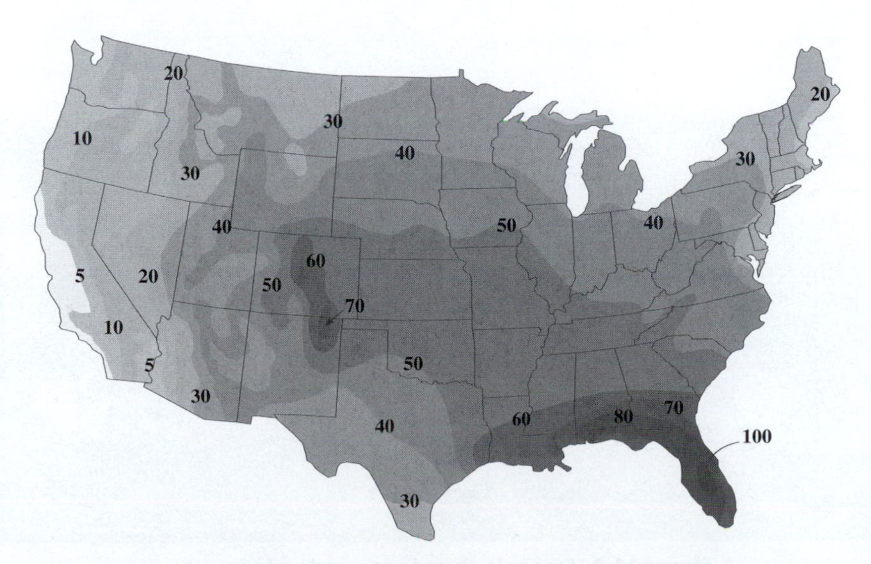

Figure 11-1. Annual average number of days with thunderstorms.

2. *What processes over the Florida peninsula force maritime tropical air to rise and lead to the high frequency of central Florida thunderstorms?*

3. *What geographic feature might contribute to the high frequency of thunderstorms in central Colorado and northeastern New Mexico?*

Lightning

Lightning is present in all thunderstorms. While it can be seen over great distances, the resulting thunder usually cannot be heard beyond 30 km. Since light travels substantially faster than sound, we see a lightning flash before we hear the accompanying thunder. In fact, we see lightning nearly instantaneously, while sound travels at approximately 340 m s^{-1} (1115 ft s^{-1}; or roughly one-fifth of a mile per second).

4. *If you see a lightning bolt in the distance and 15 seconds elapse before you hear the corresponding thunder, how far away was the lightning bolt?*

   ________ *meters*

   ________ *miles*

Air Mass Thunderstorms

Air mass thunderstorms arise when air becomes unstable within an already warm, moist air mass. This situation most commonly occurs during the warmer months, when intense surface heating creates instability within maritime tropical (mT) air masses.

We can describe the development of these thunderstorms in three stages: cumulus, mature, and dissipating. In the cumulus stage, strong updrafts dominate the circulation within a cumulonimbus cloud, causing the cloud to grow vertically. During the mature stage, precipitation begins and downdrafts develop side by side with updrafts. Cooler, drier air from outside the thunderstorm is drawn into the cloud in a process called *entrainment*. Eventually downdrafts dominate the circulation and the storm begins to dissipate. Figure 11-2 shows features of these three stages.

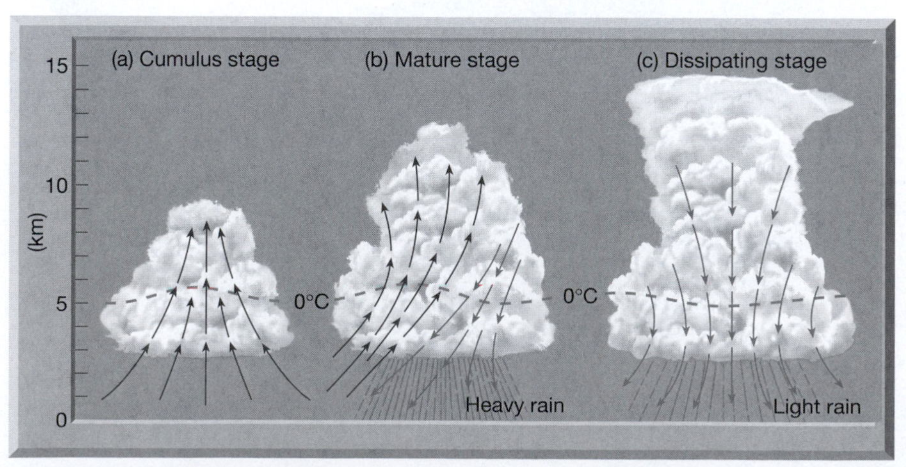

Figure 11-2. Stages in thunderstorm development.

5. *Recall that updrafts within the thunderstorm result in moist adiabatic cooling. How is energy released within the cloud in this process and how might this foster more upward motion within the cloud?*

6. *The air drawn into a thunderstorm from the outside is usually cooler than the air within a cumulonimbus cloud. Explain what effect this should have on vertical motion.*

7. *Entrained air is drier than the air within a cloud. Considering the phase change of water, how could this lead to further cooling?*

8. *How could updrafts and downdrafts contribute to cloud droplet or ice crystal growth?*

Severe Thunderstorms

Severe thunderstorms produce frequent lightning and damaging winds or hail. Like air mass thunderstorms, they need warm, moist, conditionally unstable air at the surface. In contrast with air mass thunderstorms, however, severe thunderstorms require upper-air divergence and strong vertical wind shear. These conditions often exist in association with features of mid-latitude cyclones and are very common in the central United States during the spring and early summer. At this time, the contrast between maritime tropical air from the Gulf of Mexico and continental polar air from the north is largest, and consequently the jet stream is strongest. Along a cold-front boundary, warm, moist air can be forced upward by the intruding cold air. Its ascent is fostered by the jet stream, which produces upper-air divergence and vertical wind shear.

Occasionally, a line of thunderstorms develops in the maritime tropical air mass ahead of a surface cold front. Such squall lines result from upper-air divergence associated with a strong jet stream and gravity waves triggered by the advancing cold front. The line of thunderstorms is often maintained through self-propagation. Figure 11-3 shows a cross section of a squall-line thunderstorm. It illustrates how the downdrafts of a mature thunderstorm may trigger the development of a new convective cell.

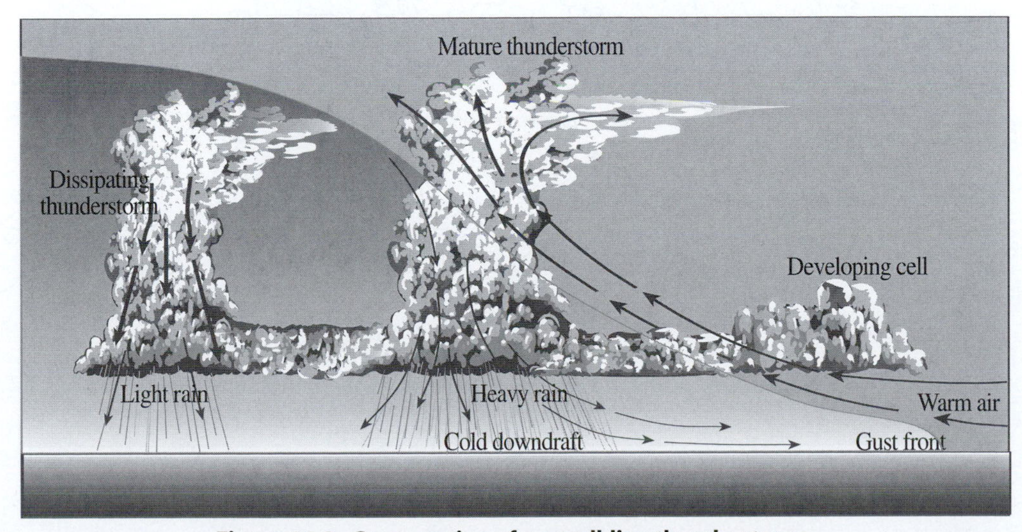

Figure 11-3. Cross section of a squall-line thunderstorm.

9. *Explain how downdrafts can trigger the development of new thunderstorm cells.*

10. *How do air mass conditions ahead of the squall line support the development of new cells?*

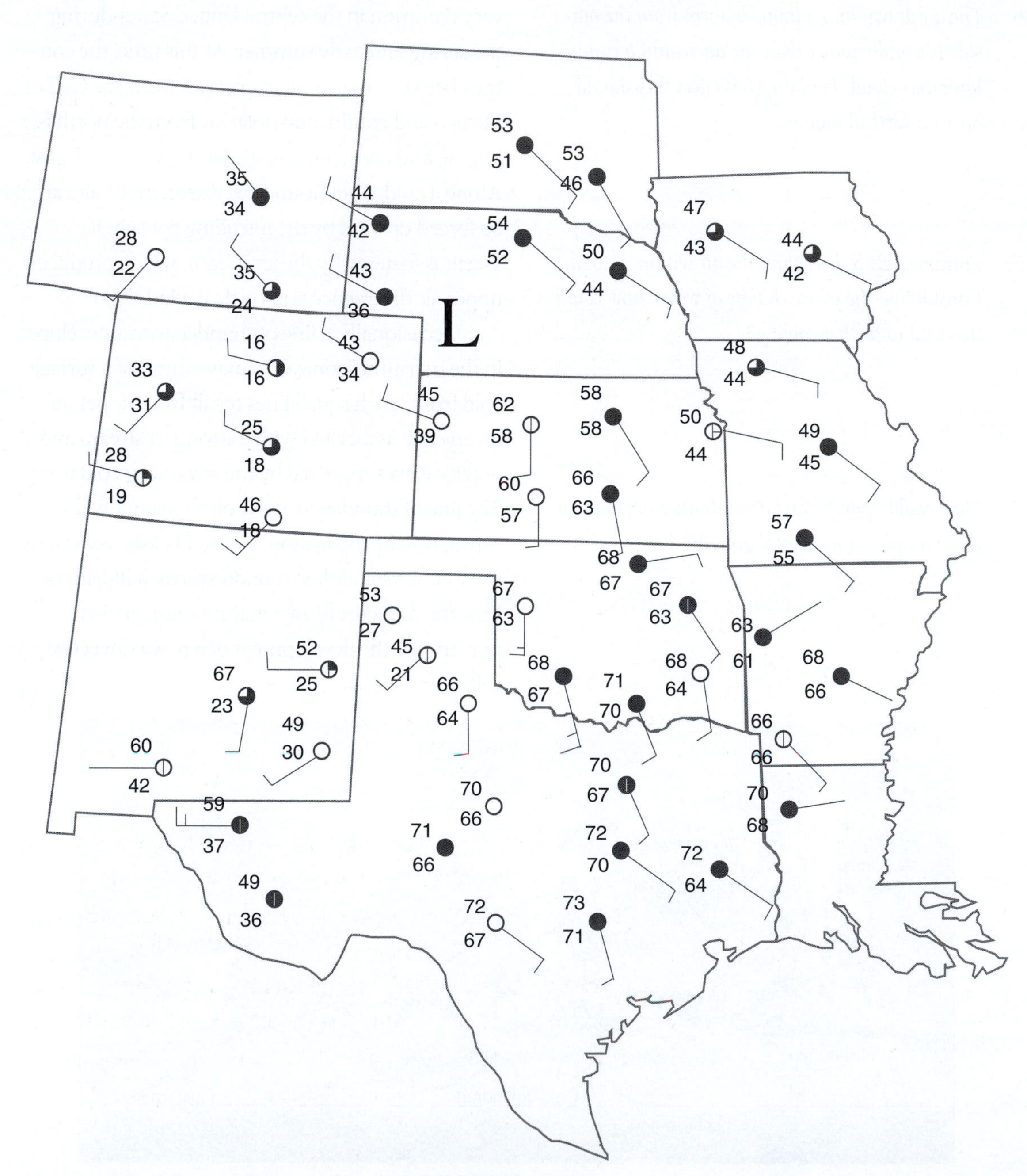

Figure 11-4

Severe thunderstorms in the United States also form along dry lines. These features develop most frequently in the Southern Plains, where continental tropical and maritime tropical air masses often meet. Because continental tropical air develops in the Southwest, where elevations are relatively high, it flows over maritime tropical air when both air masses move into the south-central United States. The continental tropical air can actually form a capping upper-air inversion that suppresses all but the strongest updrafts. This phenomenon is discussed further with respect to tornado development.

11. *Label the cP, cT, and mT air masses in the map opposite (Figure 11-4). Use the conventional symbol to draw a cold front separating the cP and mT air masses, and a dashed line to show a dry line separating cT and mT air masses.*

Tornadoes

Some severe thunderstorms produce tornadoes. The next series of exercises examines the conditions leading to severe thunderstorms and tornadoes on three separate days. First, is the tornado outbreak in Illinois, Indiana, Ohio, and Kentucky on June 2–3, 1990. Figure 11-5 shows the paths of eight tornadoes, occurring June 2, from 5:45 to 11:10 PM CDT. These tornadoes were categorized as F4 on the Fujita Intensity Scale, indicating wind speeds of 333–419 kilometers per hour (207–260 mph). They were the most devastating of 55 documented tornadoes occurring in the region that day.

Figures 11-6 through 11-9 show the surface and 500-mb maps for June 2 and 3 at 7:00 AM CDT (approximately 12 hours before and after the tornado outbreak).

Figure 11-5. F4 tornadoes on June 2, 1990.

12. *List the June 2nd surface weather conditions at two stations in the region of the tornadoes.*
    Station: ______________________
    Temperature: ______________________
    Dew point: ______________________
    Wind direction: ______________________

    Station: ______________________
    Temperature: ______________________
    Dew point: ______________________
    Wind direction: ______________________

13. *How did the surface conditions (temperature, dew point, wind direction) contribute to the day's severe weather?*

14. *Compare temperature and dew point at stations east and west of the cold front on June 2. How sharp was the contrast in air masses on this date?*

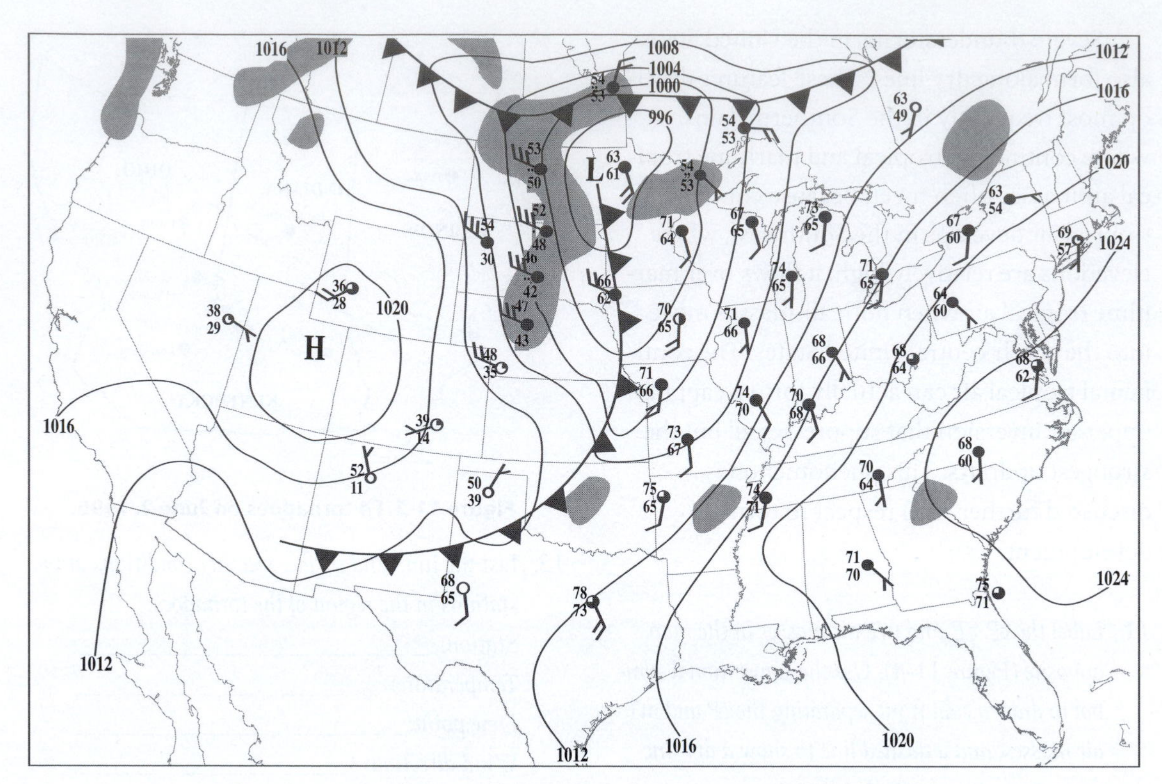

Figure 11-6. Surface conditions, Saturday, June 2, 1990, 7:00 AM CDT.

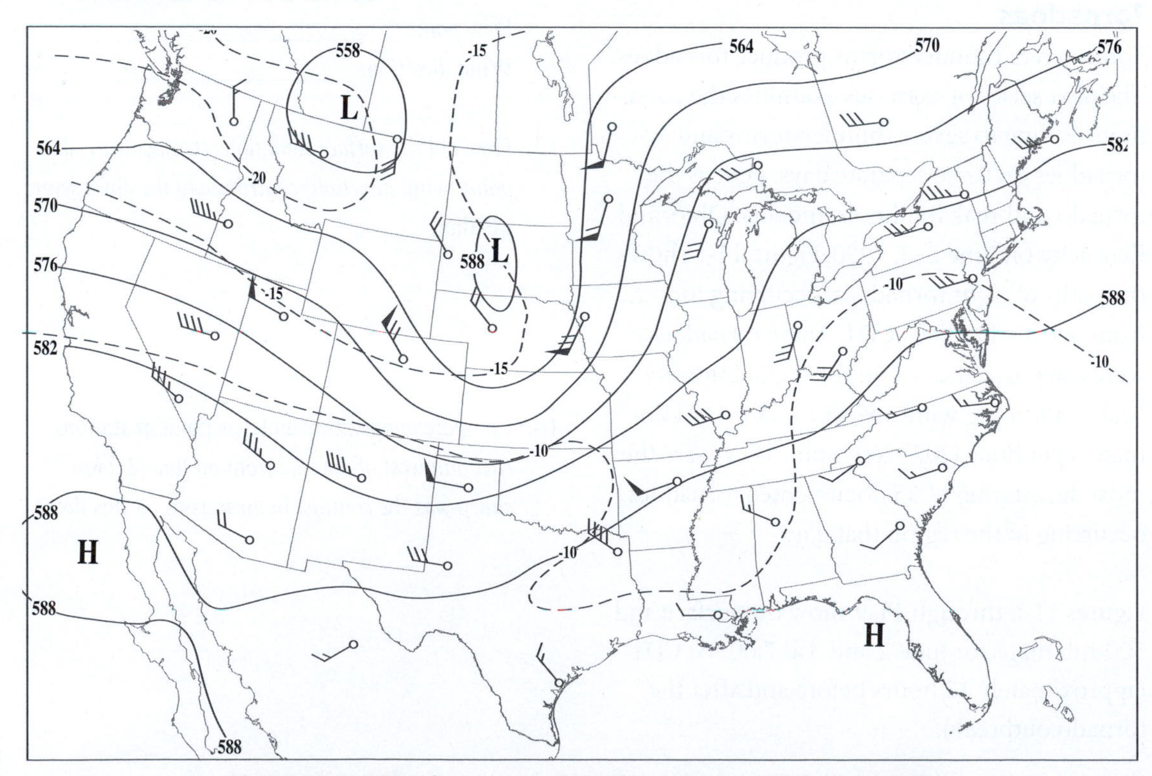

Figure 11-7. 500-mb contours, Saturday, June 2, 1990, 7:00 AM CDT.

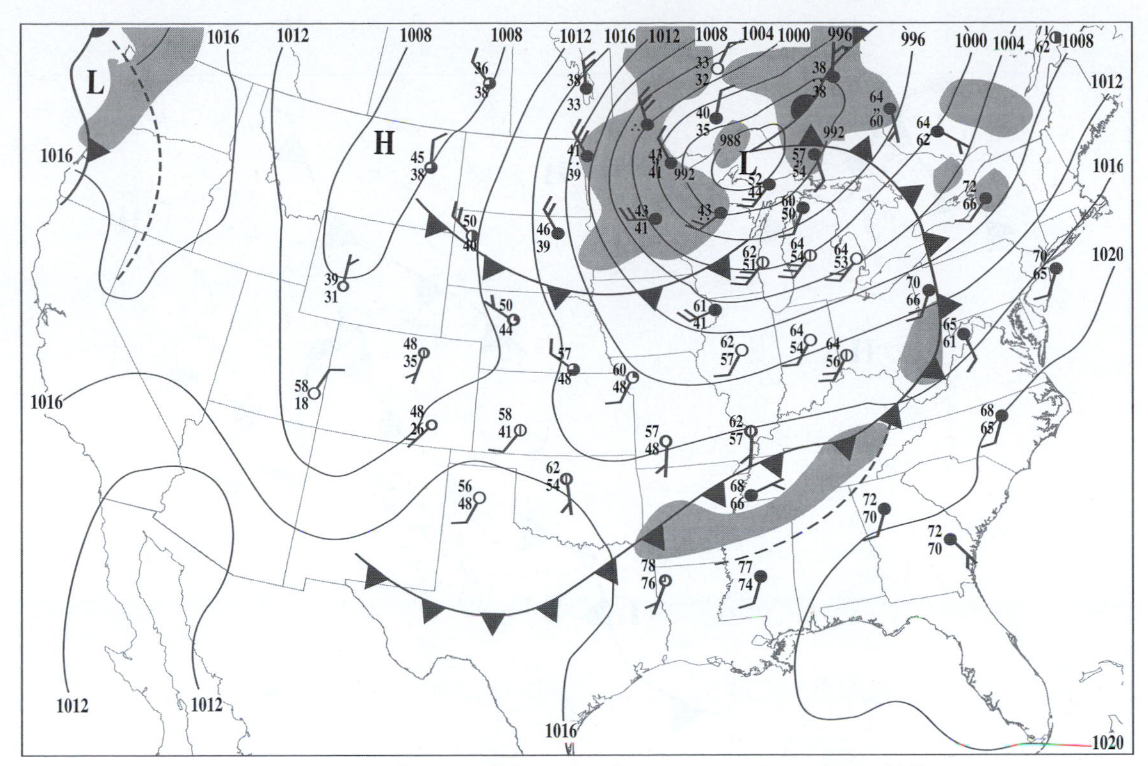

Figure 11-8. Surface conditions, Sunday, June 3, 1990, 7:00 AM CDT.

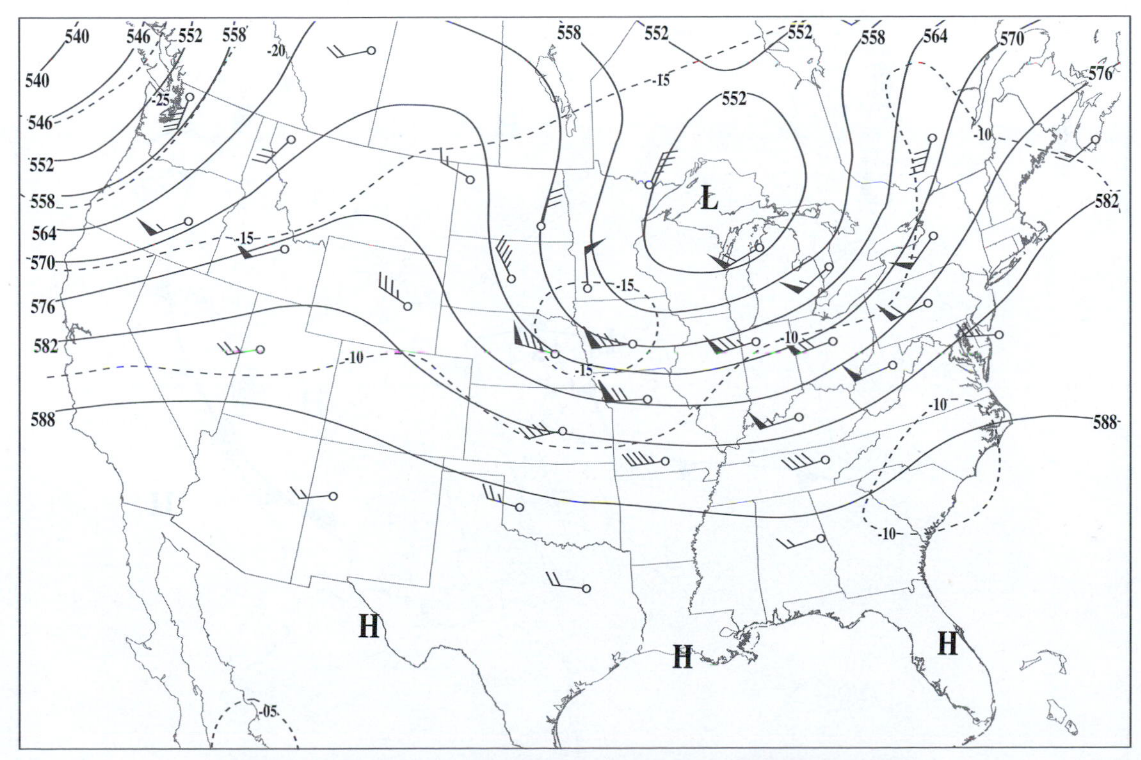

Figure 11-9. 500-mb contours, Sunday, June 3, 1990, 7:00 AM CDT.

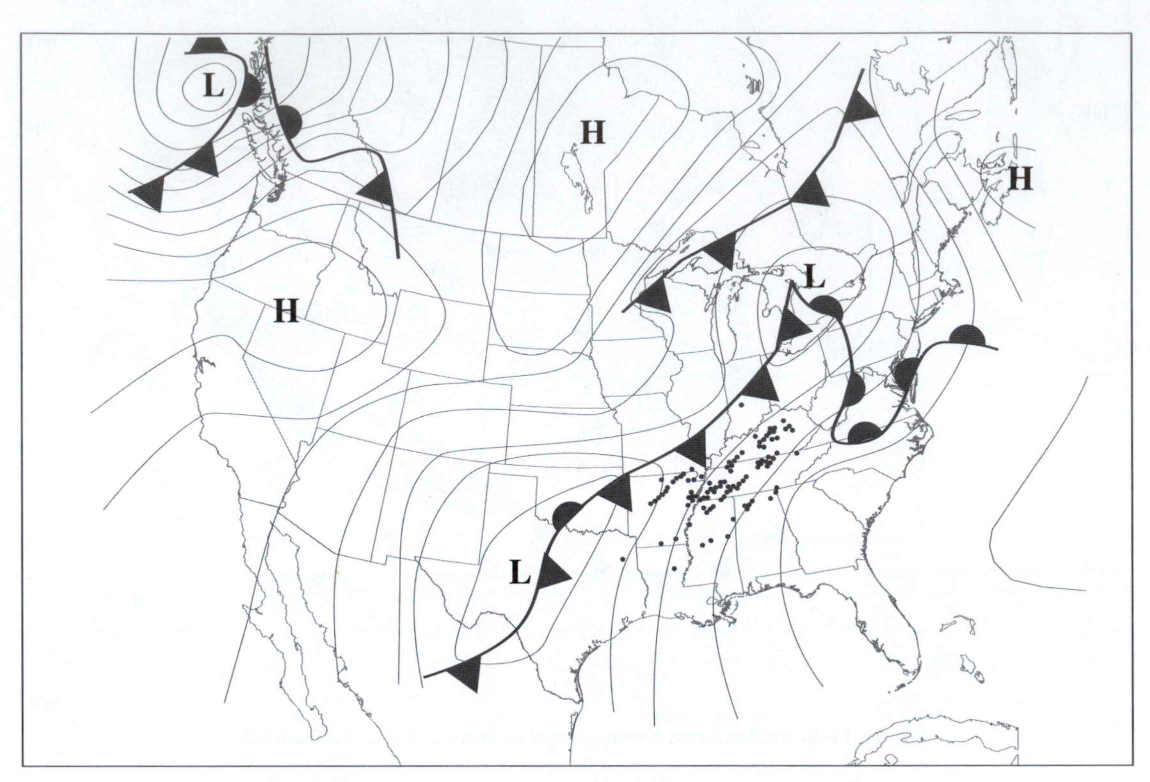

Figure 11-10. Surface pressure systems and fronts, February 5, 2008, 7:00 AM EST.

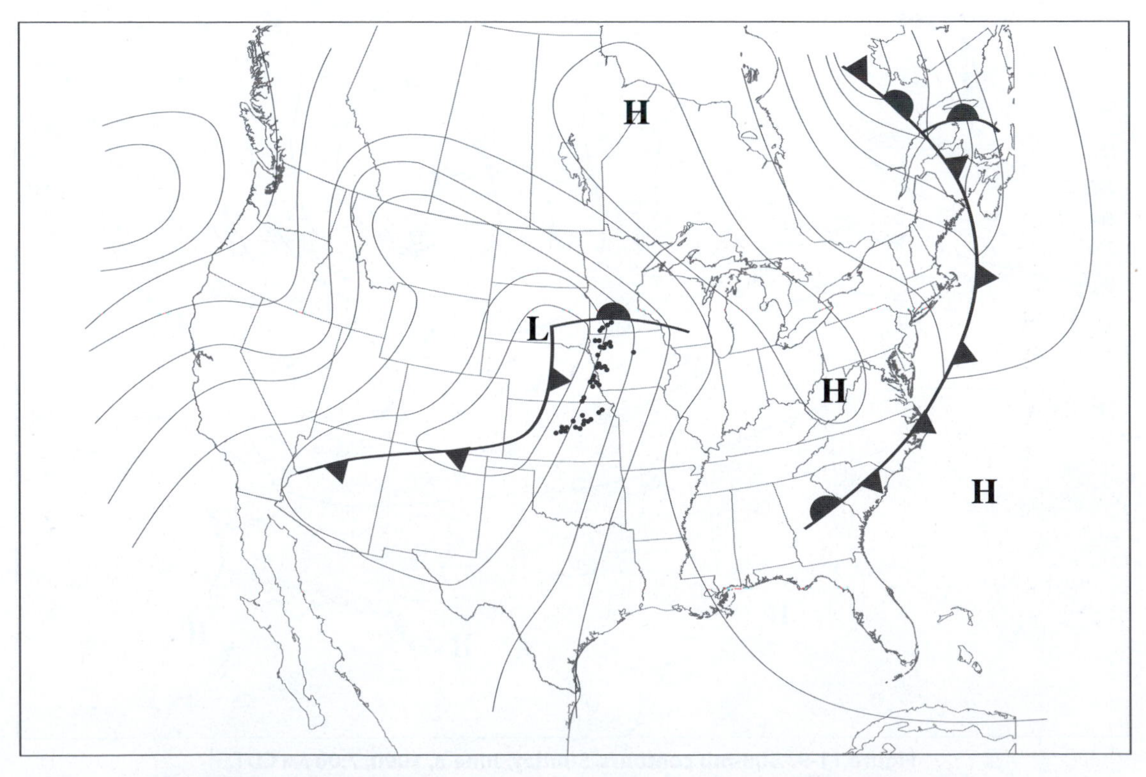

Figure 11-11. Surface pressure systems and fronts, June 11, 2008, 7:00 AM EST.

Now examine the surface weather maps during two other days with tornados. Figures 11-10 and 11-11 show the 7:00 AM EST position of surface pressure patterns and fronts on February 5, 2008 and June 11, 2008, respectively. The dots on these maps, indicate locations where tornadoes were sighted in the subsequent twenty-four hour period.

15. *What is similar about where tornadoes occur relative to the surface weather patterns on June 2-3, 1990, February 5–6, 2008, and June 11–12, 2008?*

Upper-air data at individual sites provide information that is useful for severe thunderstorm prediction. Among other clues, meteorologists examine temperature and moisture characteristics at various heights, determine the stability of the atmosphere, and search for upper-air inversions and wind shear.

An *upper-air inversion* is defined as a layer above the earth's surface in which temperature increases with height. Upper-air inversions foster severe weather by temporarily suppressing vertical mixing between warm, moist air near the surface and cool, dry air aloft. This "cap" prevents vertical cloud development. It can form when a layer of air subsides and warms adiabatically, or when continental tropical air is advected aloft; however, when triggered by a cold front or squall line, warm air can be forced into the stable inversion layer, resulting in intense convection and tall cumulonimbus clouds (Figure 11-12).

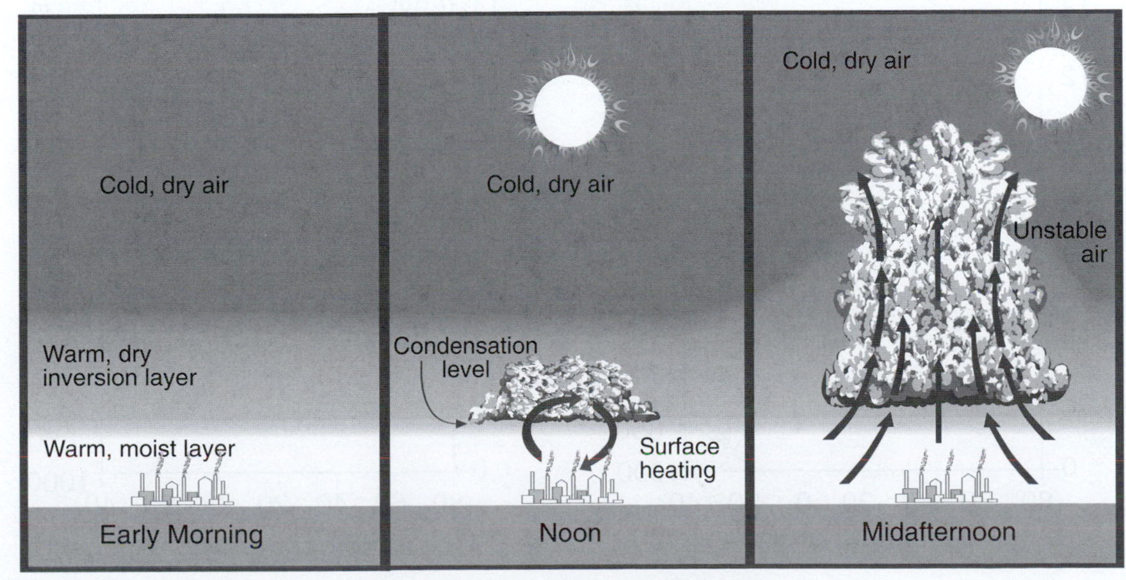

Figure 11-12. An upper-air inversion.

Data from stations near the tornado outbreaks provide an example of the upper-air conditions conducive to severe weather. Table 11-1 and Figures 11-13 through 11-15 show the environmental temperature and dew-point temperature profiles and the temperature of a rising air parcel in each environment near the time tornadoes struck.

16. *Circle the upper-air inversion layer on the Paducah environmental temperature profile graph.*

Table 11-1
Wind direction and speed during tornado outbreaks near Paducah, Little Rock, and Topeka.

	Paducah June 2, 1990, 7:00 P.M. CDT		**Little Rock** February 5, 2008, 6:00PM CST		**Topeka** June 11, 2008, 7:00 PM CDT	
Pressure (mb)	Wind Dir. (°)	Wind Speed (knots)	Wind Dir. (°)	Wind Speed (knots)	Wind Dir. (°)	Wind Speed (knots)
250	255	56	260	58	240	55
500	251	51	215	70	240	48
700	248	62	215	63	210	43
850	236	47	210	56	175	35
925	205	35	190	42	165	30

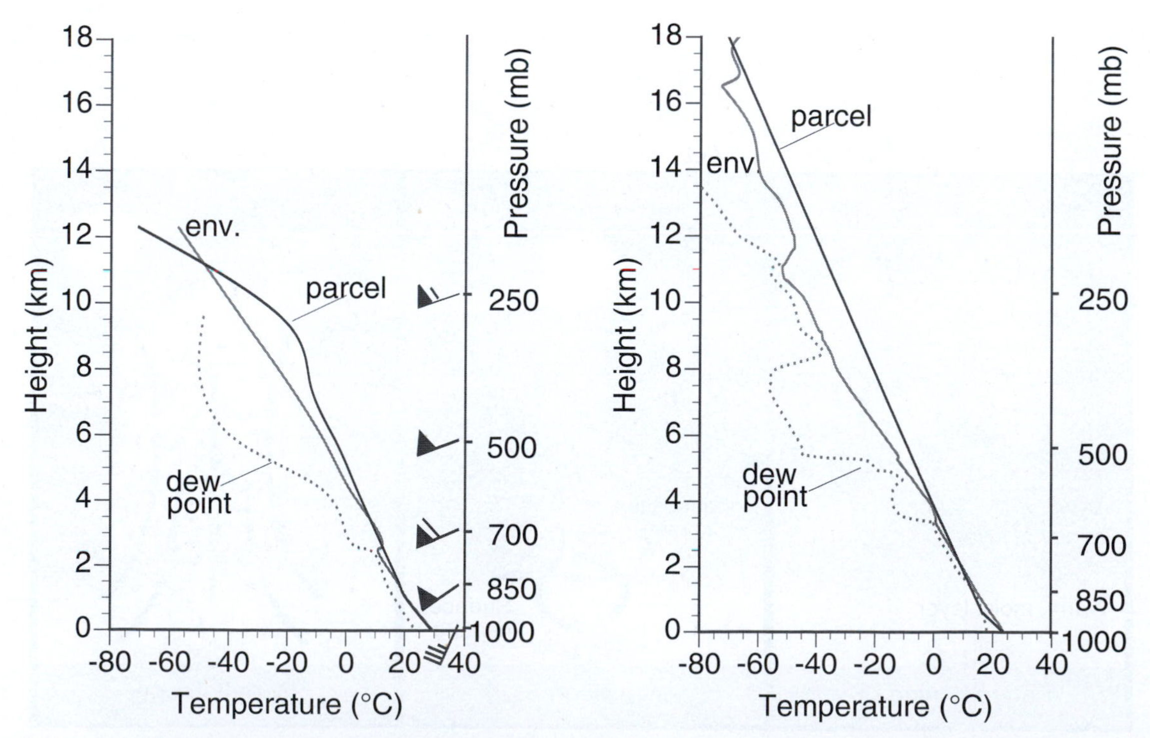

Figure 11-13. Paducah, June 2, 1990, 7:00 PM CDT. **Figure 11-14. Little Rock, February 5, 2008, 6:00 PM CST.**

17. *How would you characterize the temperature and moisture of air between the surface and 750 mb at Paducah?*

18. *How would you characterize the temperature and moisture of the air above 500 mb at Paducah?*

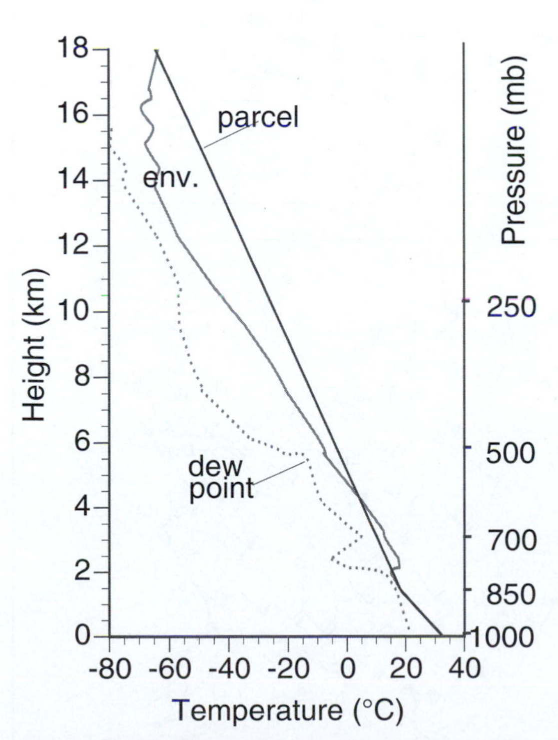

Figure 11-15. Topeka, June 11, 2008, 7:00 PM CDT.

Vertical Wind Shear

Table 11-1 lists the wind speed and direction at various heights for the same date and time of each temperature profile. To interpret these values, consider a circle of 360°. North is at the top of the circle (0° or 360°), east is on the right (90°), south is at the bottom of the circle (180°), and west is on the left (270°). For example, Paducah, Kentucky, had a west-southwest wind (250°) at 250 mb.

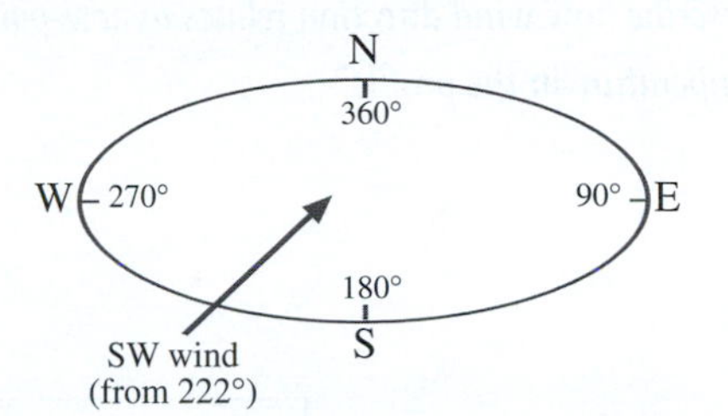

Figure 11-16

Wind speed and direction are important because the temperature and moisture characteristics of air advected into a region can determine its stability. When wind direction changes with height, advection can bring warm, moist air into certain layers of the atmosphere and cool, dry air into others.

Vertical changes in wind speed and direction influence the severity of thunderstorms. Air mass thunderstorms develop when winds are light throughout the troposphere. Severe thunderstorms form when wind speed and direction change rapidly with height. Some severe thunderstorms will be multicell squall-line storms; others may develop into rotating supercell thunderstorms. The formation of supercell thunderstorms, which produce the most violent tornadoes, requires strong winds just above the surface that increase in speed and veer (turn clockwise) with altitude.

19. *Using the wind speed and direction symbols in Figure 11-13 (Paducah) as a guide, draw the wind speed and direction symbols at 250, 500, 700, 850, and 925 mb on Figures 11-14 and 11-15.*

Figures 11-17 through 11-21 provide a more thorough review of wind shear present during the February 5-6 tornado outbreak.

20. *How do the winds at 925-mb and 850-mb differ from those at higher levels?*

21. *Describe how wind direction relates to dew-point temperature in the profile?*

22. *How do the winds at each level contribute to severe weather on February 5-6?*

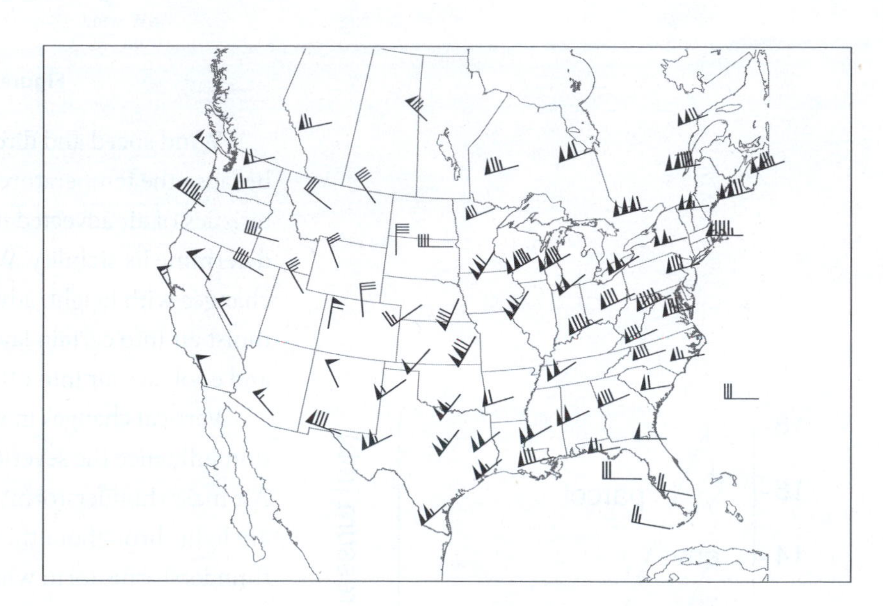

Figure 11-17. 250-mb winds, February 5, 2008, 6:00 PM CST.

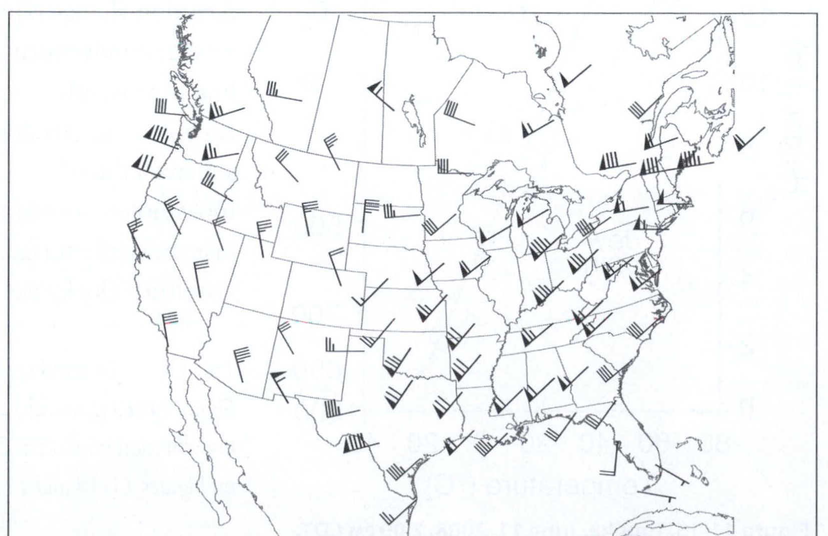

Figure 11-18. 500-mb winds, February 5, 2008, 6:00 PM CST.

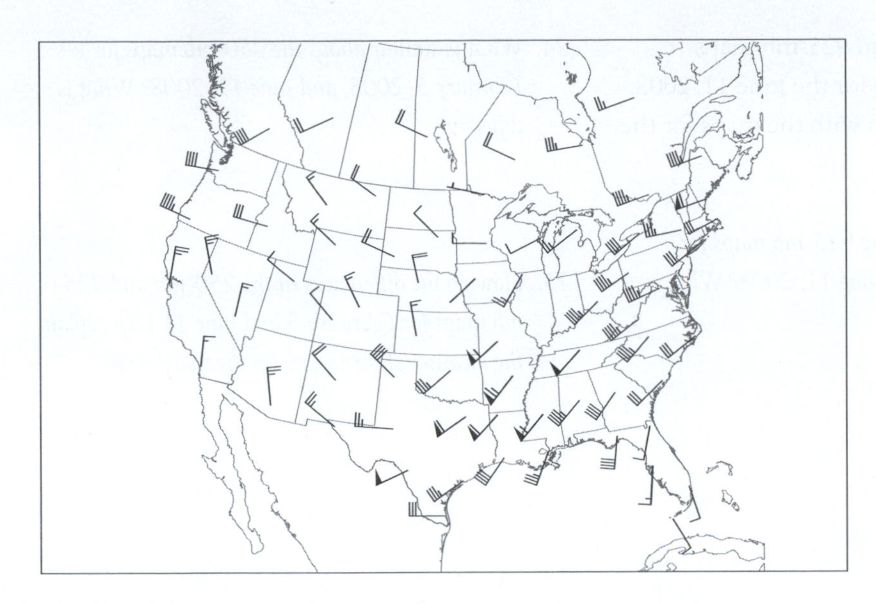

Figure 11-19. 700-mb winds, February 5, 2008, 6:00 PM CST.

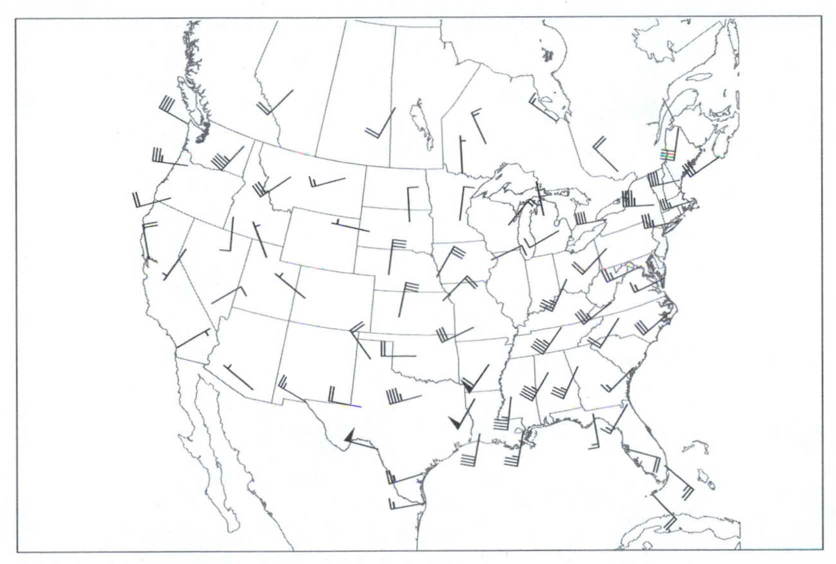

Figure 11-20. 850-mb winds, February 5, 2008, 6:00 PM CST.

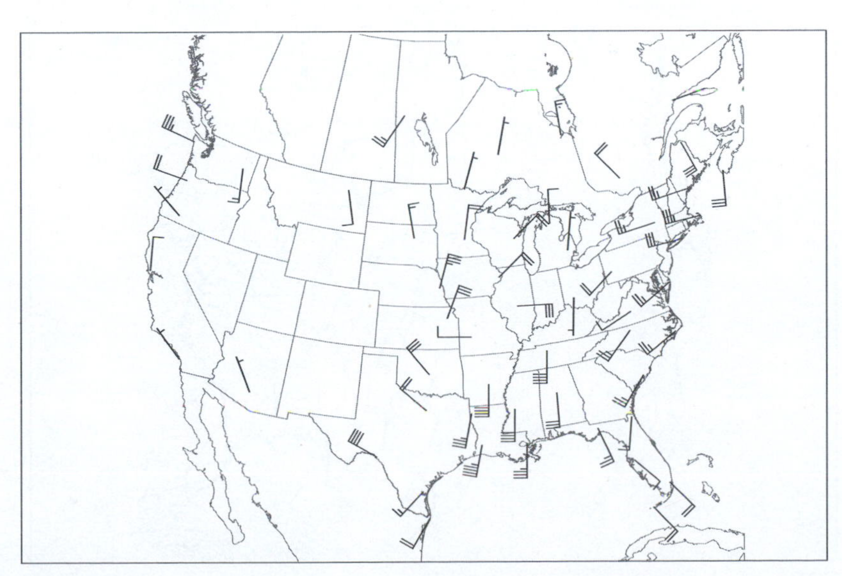

Figure 11-21. 925-mb winds, February 5, 2008, 6:00 PM CST.

Now examine the 250- and 925-mb maps (Figures 11-22 and 11-23) for the June 11, 2008, storm, and compare them with the maps for the February storm.

23. *What is similar about the 925-mb maps for February 5, 2008, and June 11, 2008? What is different?*

24. *What is similar about the 250-mb maps for February 5, 2008, and June 11, 2008? What is different?*

25. *How do the differences in the 250-mb and 925-mb maps for February 5 and June 11 help explain the location of tornadoes on the two dates?*

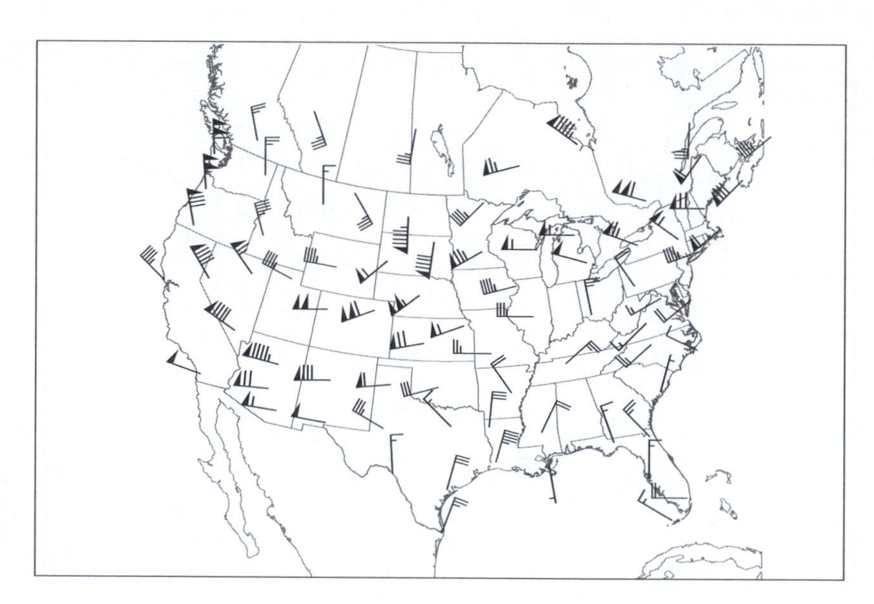

Figure 11-22. 250-mb winds, June 11, 2008, 7:00 PM CDT.

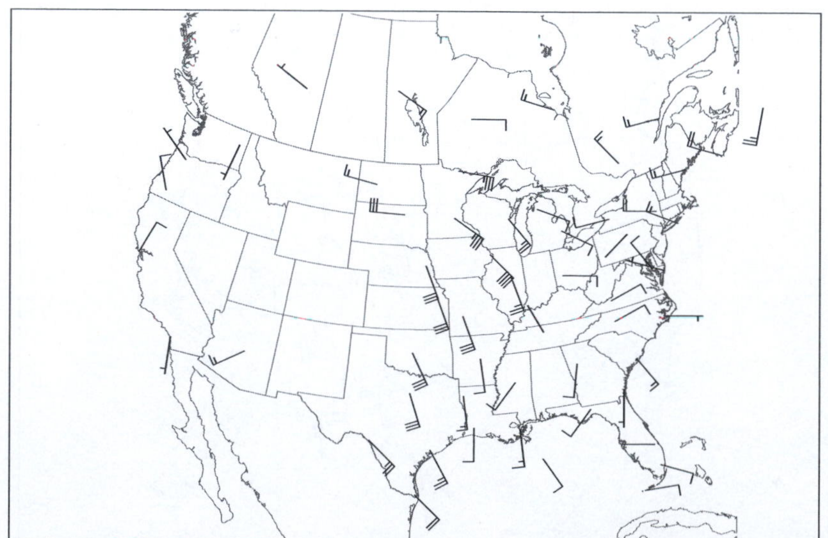

Figure 11-23. 925-mb winds, June 11, 2008, 7:00 PM CDT.

Review Questions

Describe the major processes associated with the mature stage of a thunderstorm.

What surface weather patterns lead meteorologists to predict severe thunderstorms?

What upper-air patterns would support a forecast for severe thunderstorms or tornadoes?

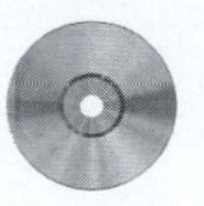

Lab 12

HURRICANES

Introduction

Tropical cyclones are violent storms characterized by a low pressure center, high winds, heavy rain, and rough seas. They form over tropical oceans and are fueled by the evaporation of warm water and conversion of thermal energy into kinetic energy. Those that form in the Atlantic and eastern Pacific are commonly called hurricanes. This lab examines the structure, energy, and movement of hurricanes and uses Hurricanes Hugo and Katrina as case studies to illustrate these concepts.

Circulation within a Hurricane

Air circulates around and in toward a hurricane's low pressure center—the *eye* of the storm. Surface wind speeds are light in the eye, reaching a maximum approximately 25 km from the eye (an area called the *eye wall* because of its tall cumulonimbus clouds), and diminish with distance from the eye wall.

1. *Draw the surface winds around the eye of the Northern Hemisphere hurricane depicted in Figure 12-1.*

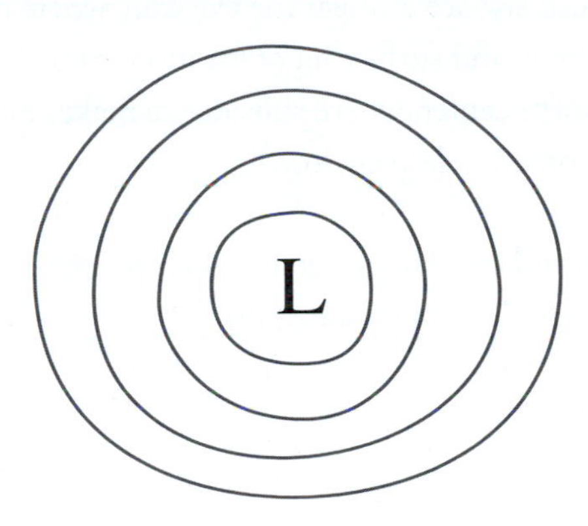

Figure 12-1

Figure 12-2 shows a cross-sectional view of the horizontal and vertical motion within hurricanes. As air moves toward the eye the storm gains thermal energy through evaporation from the warm ocean surface. Air rises near the storm's center, forming tall cumulonimbus clouds that constitute the eye wall. At high levels of the troposphere, air diverges (moves away) from the central eye, an area of weak high pressure. This air loses energy through long-wave radiation emission, cools, and subsides at the dry adiabatic lapse rate several hundred kilometers from the eye. Some of the upper-level air sinks to the surface, where it flows back toward the eye. In addition, subsidence in the eye itself reduces cloud cover.

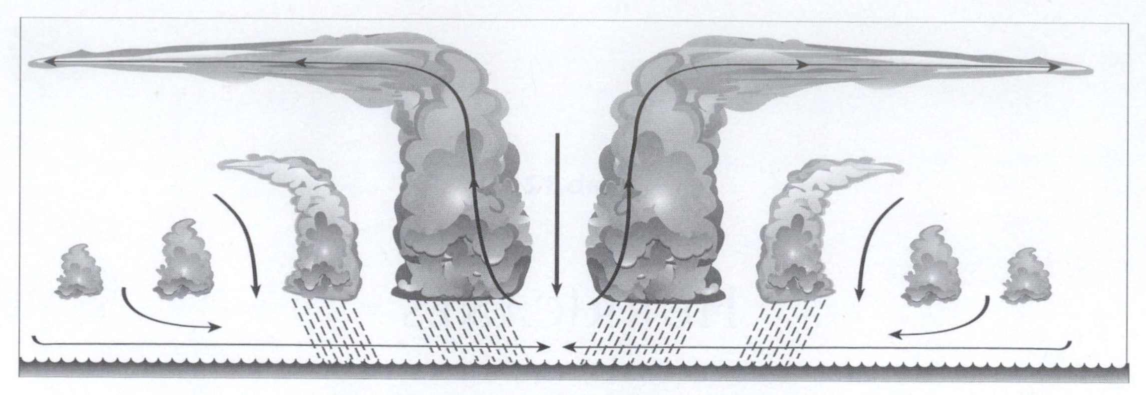

Figure 12-2. Circulation within a hurricane.

Just above the tropical waters where hurricanes form, air is uniformly warm. However, moisture content of the air varies considerably between surface air near the eye wall, where it is very high, and surface air at locations away from the storm center, where subsidence makes moisture content relatively low.

2. *How does subsidence at the storm's periphery make surface air relatively dry?*

3. *Hurricanes gain their energy through evaporation of warm ocean waters. How does the subsidence of dry air affect evaporation rate? How does wind speed affect evaporation rate?*

4. *How do sea-surface temperatures affect evaporation rate? What could cause variations in sea-surface temperatures ahead of a hurricane's path?*

Hurricane Hugo

Hurricane Hugo began as a cluster of thunderstorms off the West African coast on September 9, 1989. Over the following two weeks it developed into a strong hurricane, striking Guadeloupe, the U.S. Virgin Islands, and Puerto Rico before making landfall on the U.S. coast. The storm caused a total of 86 deaths; its 20-ft storm surge and severe winds contributed to damage exceeding $9 billion. We will examine this storm to learn about the paths of hurricanes, as well as their destructive winds and storm surge.

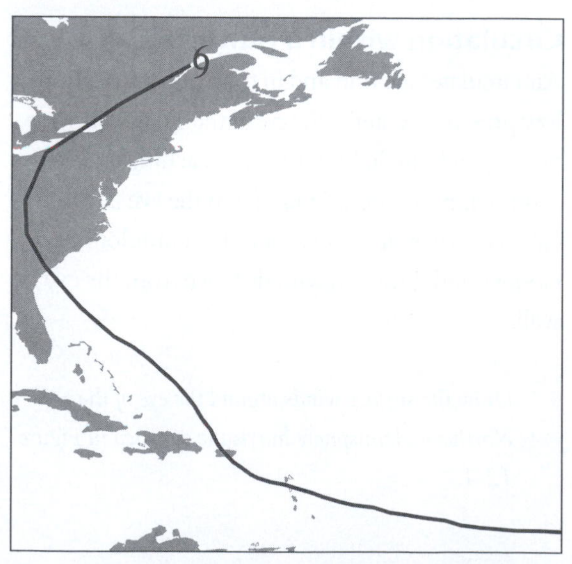

Figure 12-3. The path of Hurricane Hugo, September 1989.

Figure 12-4 shows the 700-mb wind speed at four locations when Hugo struck the South Carolina coast.

5. *On the figure draw an arrow showing the forward movement of the storm, perpendicular to the coast.*

6. *How might the forward movement of the storm influence the wind speed within the storm as depicted at the four locations below?*

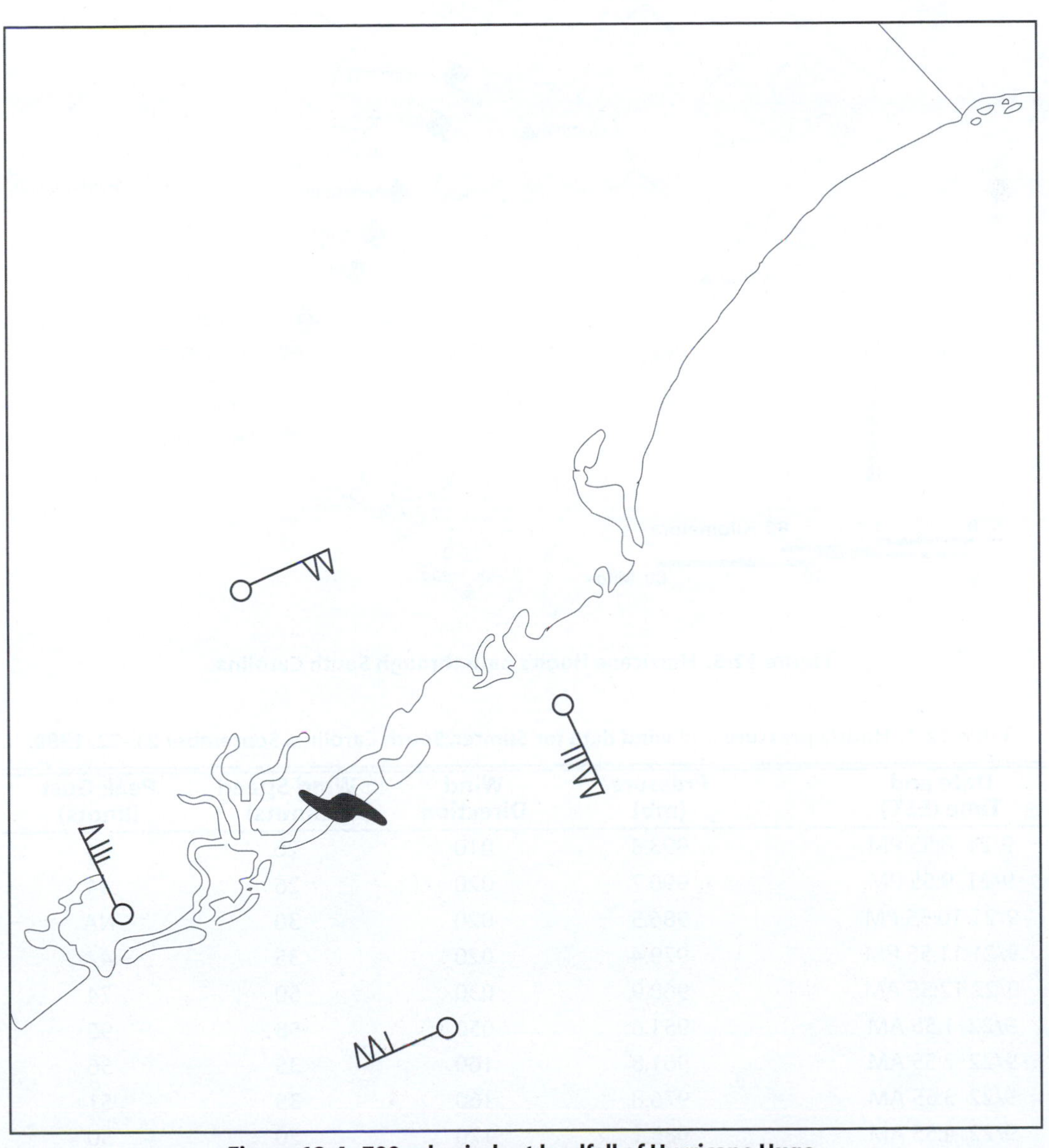

Figure 12-4. 700-mb winds at landfall of Hurricane Hugo.

Figure 12-5 shows Hugo's path through South Carolina. It struck the coast on September 21 at 11:05 PM (2305 EST) and left the state southwest of Charlotte, North Carolina, at 5:40 AM September 22.

7. *Table 12-1 lists hourly pressure and wind data for Sumter, South Carolina. Explain the drastic change in wind direction at Sumter that occurred between 12:55 AM and 2:55 AM.*

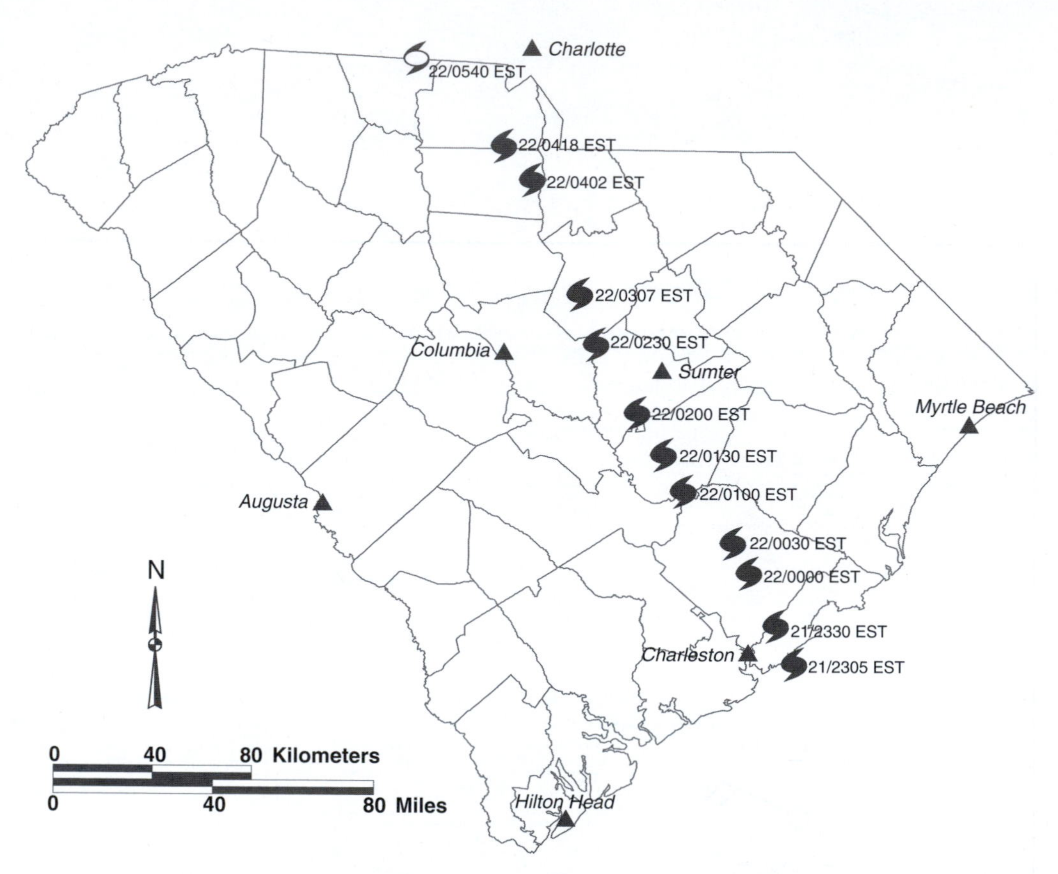

Figure 12-5. Hurricane Hugo's path through South Carolina.

Table 12-1. Hourly pressure and wind data for Sumter, South Carolina, September 21–22, 1989.

Date and Time (EST)	Pressure (mb)	Wind Direction	Wind Speed (knots)	Peak Gust (knots)
9/21 8:55 PM	993.3	010	26	34
9/21 9:55 PM	990.7	020	26	34
9/21 10:55 PM	986.5	020	30	NA
9/21 11:55 PM	979.4	020	35	47
9/22 12:55 AM	968.9	030	50	74
9/22 1:55 AM	951.6	050	58	95
9/22 2:55 AM	961.8	160	35	56
9/22 3:55 AM	976.8	160	35	51
9/22 4:55 AM	985.1	170	20	50
9/22 5:55 AM	990.7	170	24	38

Table 12-2. Storm surge.

Location	Feet
North Myrtle Beach	11.1
Myrtle Beach	13.9
Surfside Beach	12.9
Murrells Inlet	12.6
Pawleys Island Beach	12.8
North Island	8.2
Santee Point	12.1
McClellanville	16.4
Bull's Bay	20.2
Bull Island	16.2
Isle of Palms	16.2
Sullivans Island	15.8
Charleston	10.9
Folly Beach	12.1
South Kiawah Island	10.6
Rockville	5.7

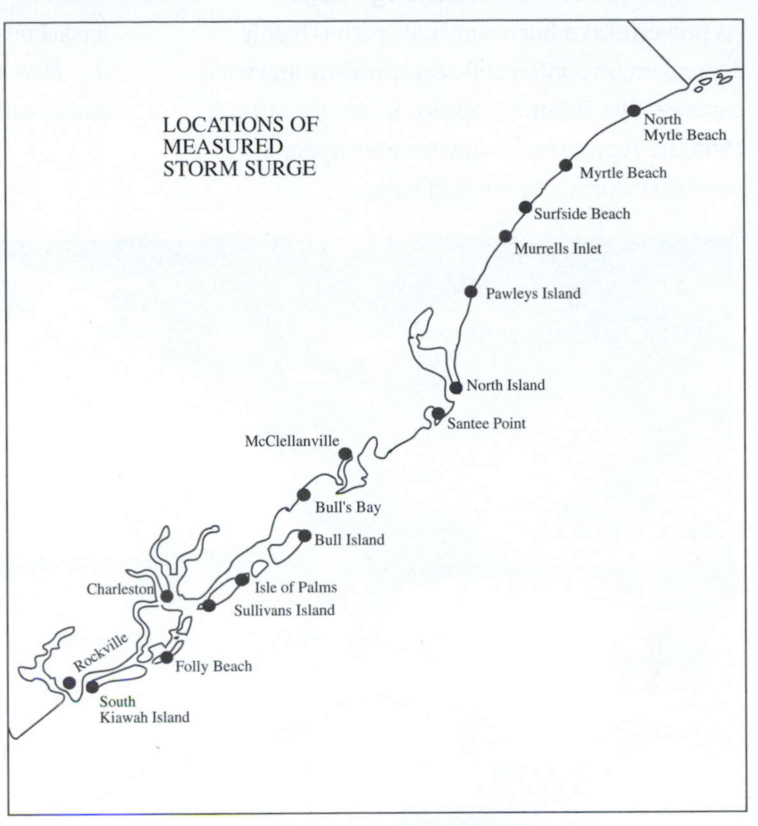

Figure 12-6. Locations of measured storm surge along the South Carolina coast.

Hurricanes create a storm surge—a dome of water as wide as 100 km—formed because of low pressure and, more importantly, strong winds that "push" water forward. When this dome approaches a coast, water "piles up," causing extensive coastal flooding. Shortly after Hurricane Hugo struck, the United States Geological Survey measured evidence of Hugo's storm surge and compiled the data seen in Table 12-2. (For reference, notice Charleston and Myrtle Beach on both Figure 12-5 and 12-6).

8. *Mark the areas of highest storm surge in Figure 12-6. How did the path of the storm influence the storm-surge pattern?*

9. *How might the configuration of the coast influence storm-surge height?*

10. *How could the timing of a hurricane landfall on a particular day influence storm-surge height?*

General Circulation during Hugo

As powerful as a hurricane is, its path is highly dependent on continental-scale pressure and wind patterns. The 500-mb maps for September 19–22, 1989, are reproduced below to show upper-air wind flow during the week of Hugo.

11. *Assuming gradient wind flow, draw the winds around the high and low pressure centers for each day. How did the circulation around these pressure centers influence Hugo's path?*

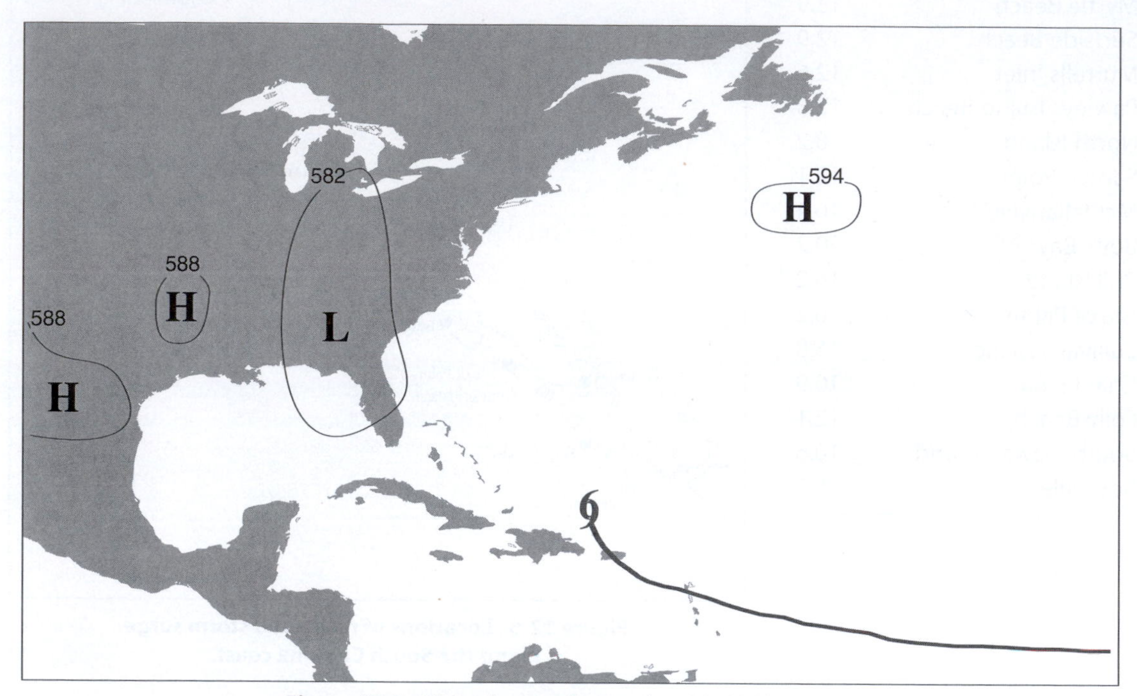

Figure 12-7. 500-mb map, September 19, 1989, 1200z.

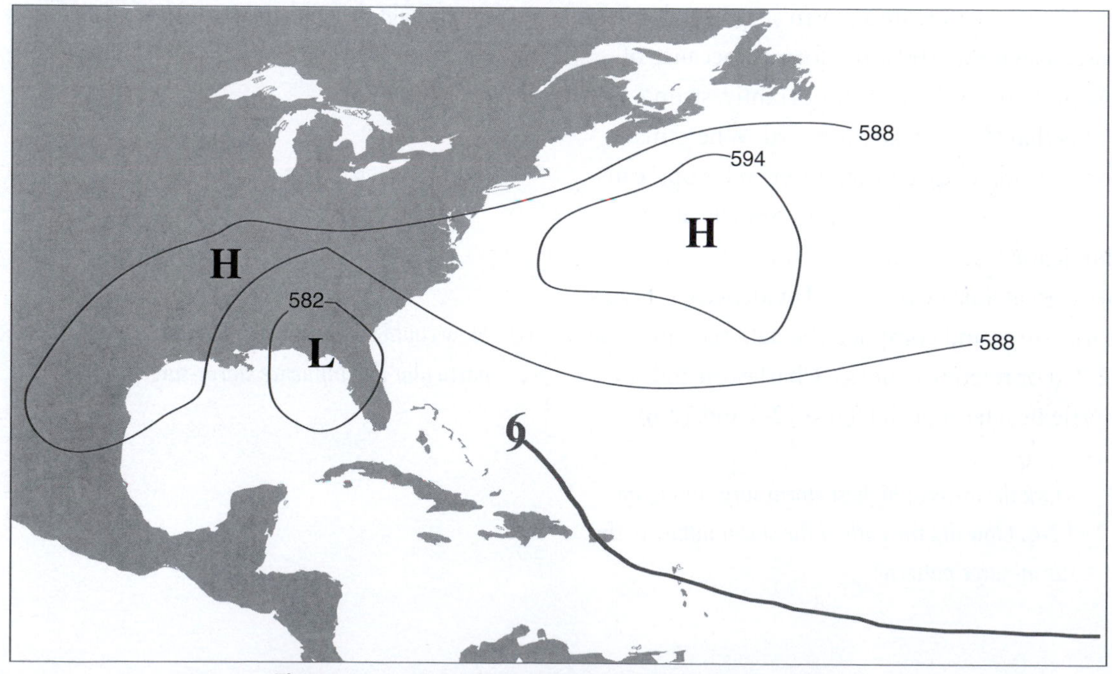

Figure 12-8. 500-mb map, September 20, 1989, 1200z.

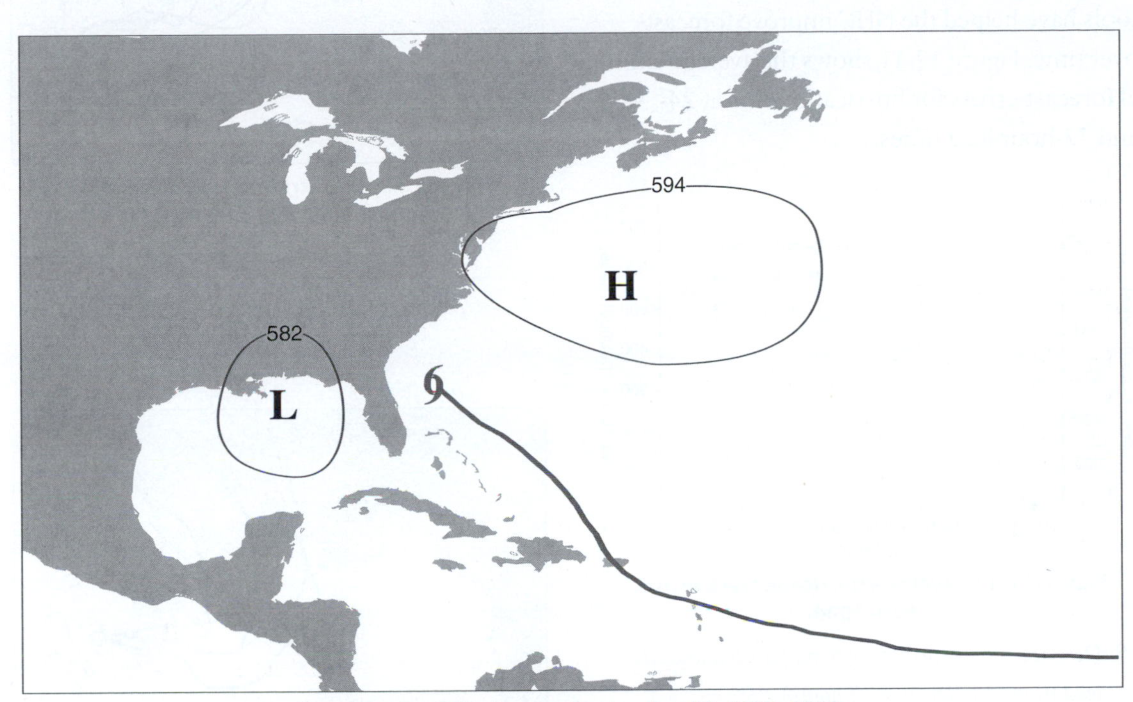

Figure 12-9. 500-mb map, September 21, 1989, 1200z.

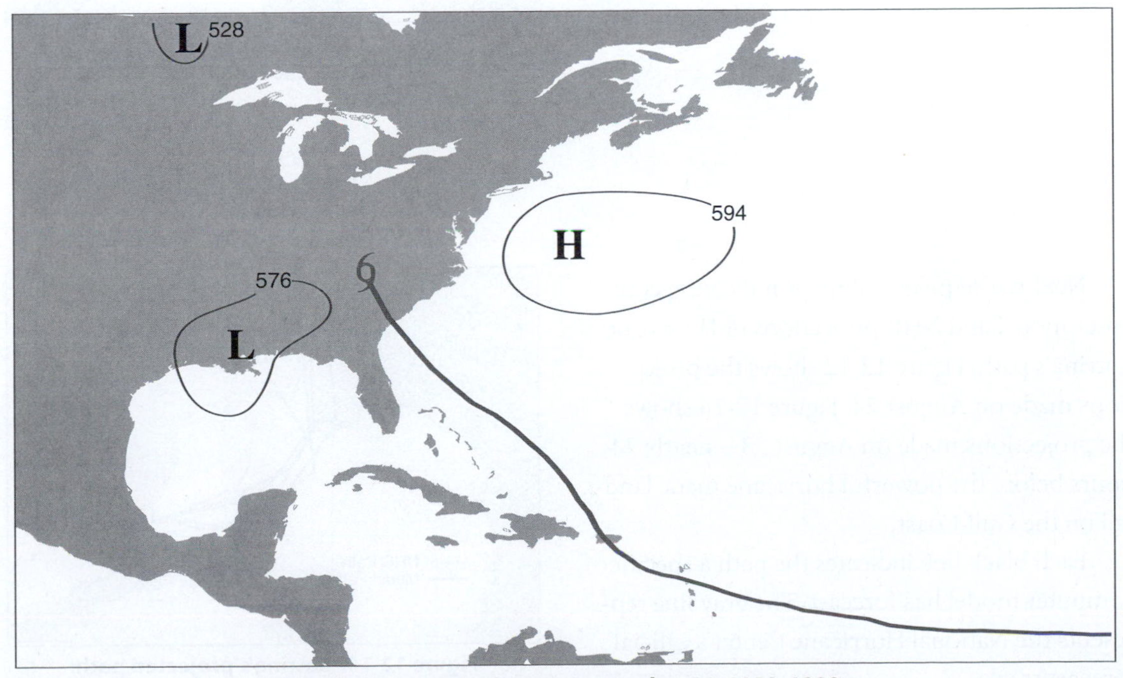

Figure 12-10. 500-mb map, September 22, 1989, 1200z.

Hurricane Forecasting

The National Hurricane Center (NHC) issues official hurricane forecasts, watches, and warnings on the basis of data from reconnaissance planes, buoys, satellites, and radar and from numerical models of the atmosphere. These tools have helped the NHC improve forecasts over time. Figure 12-11 shows the average annual forecast errors for hurricane paths at 24-, 48-, and 72-hour lead times.

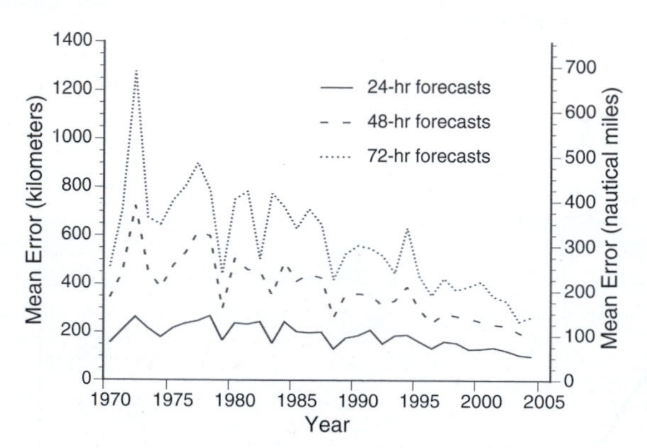

Figure 12-11. Average hurricane track errors, 1970–2004.

12. *Describe the historic error trend for each forecast lead time (24, 48, and 72 hours).*

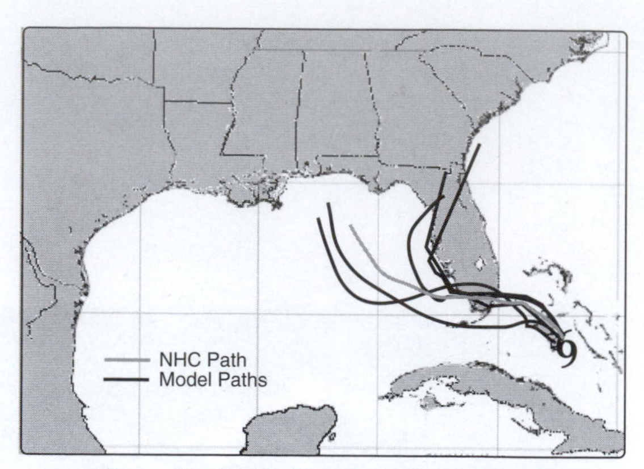

Figure 12-12. Katrina's projected path, August 24, 2005, 8:00 AM EST.

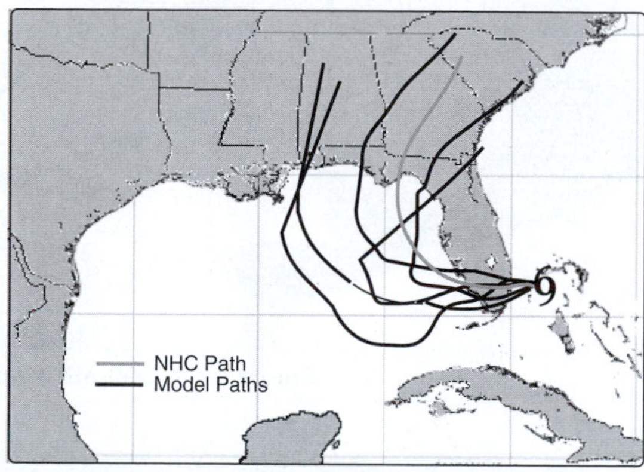

Figure 12-13. Katrina's projected path, August 25, 2005, 8:00 AM EST.

Next is a sequence of maps indicating computer model and NHC projections of Hurricane Katrina's path. Figure 12-12 shows the projections made on August 24; Figure 12-16 shows the projections made on August 28—nearly 24 hours before the powerful hurricane made landfall on the Gulf Coast.

Each black line indicates the path a specific computer model has forecast. The gray line represents the National Hurricane Center's official forecast track.

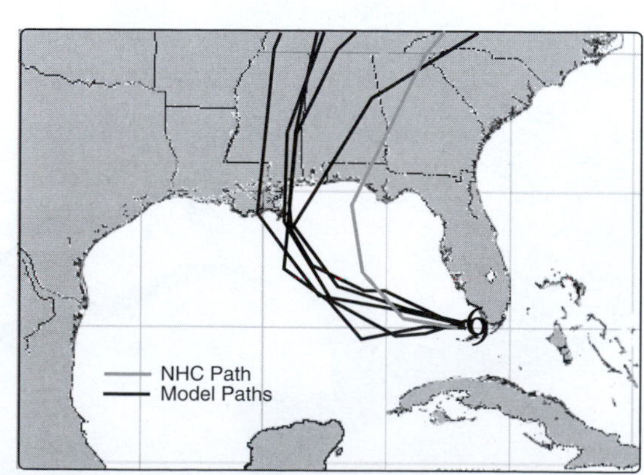

Figure 12-14. Katrina's projected path, August 26, 2005, 8:00 AM EST.

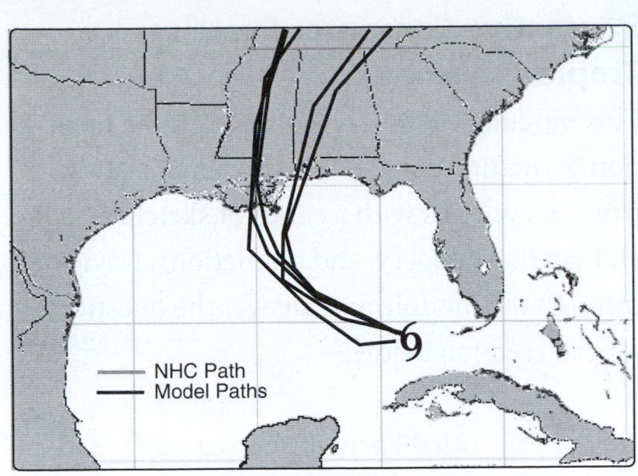

Figure 12-15. Katrina's projected path, August 27, 2005, 8:00 AM EST.

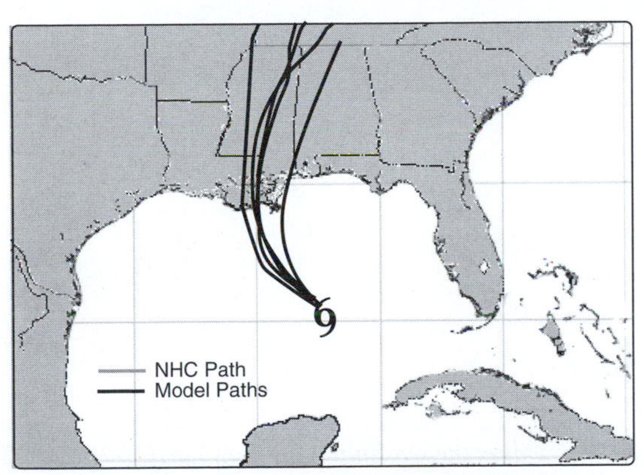

Figure 12-16. Katrina's projected path, August 28, 2005, 8:00 AM EST.

13. *How would you characterize the relative range of paths forecasted by the various computer models on August 25?*

14. *How would you characterize the range of paths projected on August 27?*

15. *Between which two consecutive days do you see the most dramatic shift in model-predicted paths?*

The National Hurricane Center issues a hurricane warning for a coastal region when they expect it to experience hurricane-force winds within 24 hours. Although the size of the warning area varies depending on storm intensity and size, population density, and the orientation of the coast relative to the storm movement, the average hurricane warning area extends approximately 556 km (300 nautical miles). The average damage swath is typically smaller than the warning area because of uncertainties in a hurricane's future path. When issuing a warning, the NHC balances caution with a desire to reduce "false alarms"—times when potential disaster does not materialize.

16. *Consider a hurricane that creates a swath of damage 185 km (100 nautical miles) wide, completely within an average-sized warning area. What fraction of the warning area would have had a false alarm?*

17. During the period 1995–2004, the National Hurricane Center's average 24-hour forecast had an approximate error of 140 km (75 nautical miles). At 48 hours, this 10-year average is approximately 255 km (138 nautical miles). Examine the 24-hour and 48-hour NHC forecasts and the actual path in the maps below. How did the NHC prediction for this storm fare relative to these average errors?

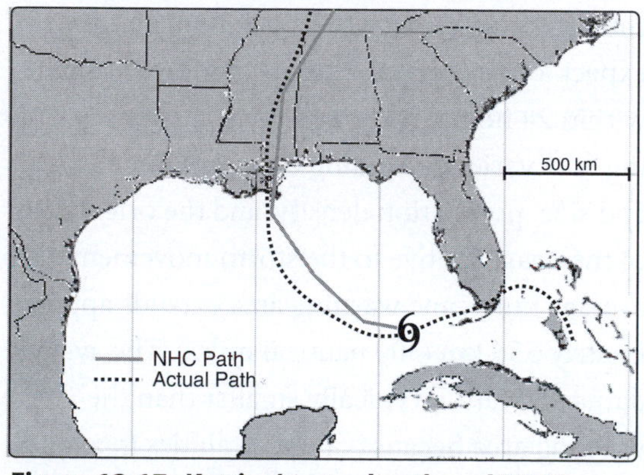

Figure 12-17. Katrina's actual path and NHC's projected path 48 hours before landfall.

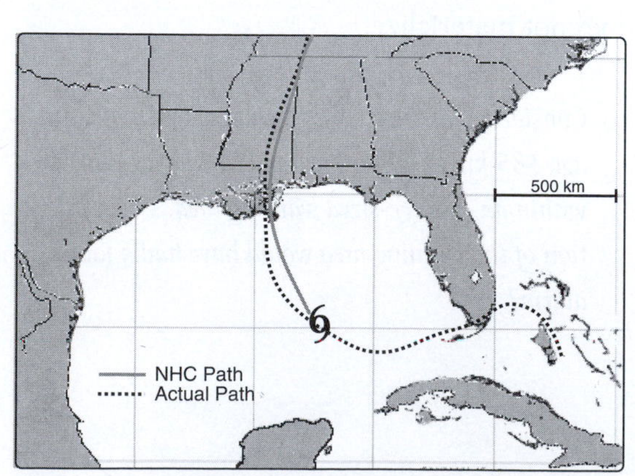

Figure 12-18. Katrina's actual path and NHC's projected path 24 hours before landfall.

Interactive Computer Exercise: Tropical Cyclones

This module will guide you through the formation, structure, energy, and movement of tropical cyclones with a series of sketches, radar and satellite imagery, and animations. Navigate through the module and answer the questions for each section below.

On the page titled Formation : Sea-surface Temperatures:

18. Briefly describe how sea-surface temperatures change between the beginning, middle, and end of the Atlantic hurricane season. What accounts for these seasonal changes?

19. How does the seasonal pattern of hurricane origin points change from the early season (June and early July), to midseason (late August and early September), to the late season (late October and November), and how do these changes relate to sea-surface temperatures?

On the page titled Formation : Easterly Wave Animation:

20. *Describe how the size, structure, and position of the animated easterly wave changes as it moves.*

On the page titled Structure : Surface Conditions:

21. *Click on the red dots showing the Hurricane Georges (1998) path. With each click you will see the meteorological conditions recorded at buoys in the Gulf of Mexico. Choose a time segment of three consecutive dots and explain the changes in meteorological conditions that occur at the buoys.*

On the page titled Energy : Cross Section:

22. *Examine the cross-sectional sketch of a hurricane by rolling over each letter. At the ocean surface, how big a temperature difference exists between the eye and the outer periphery of the storm?*

23. *The energy efficiency of a hurricane depends on the temperature difference between those areas where heat is added (ocean surface) and where it is lost (cloud tops). How different are temperatures between the sea surface and the cloud tops?*

On the page titled Energy : Landfall:

24. *Examine the animation of Hurricane Georges making landfall in Louisiana. How do its features change once it reaches land? Why?*

On the page titled Path : Past Tracks:

25. *Examine the past tracks of Atlantic hurricanes shown in the first page of the module's "Path" section. Discuss the general circulation features that influence the paths of the hurricanes shown in this graphic.*

On the page titled Path : Tracking Katrina:

26. *At what point does the National Hurricane Center report that Katrina would likely develop into a major hurricane? Why did they come to this conclusion?*

Review Questions

What fuels a hurricane?

Describe wind speed and direction within a hurricane in the Southern Hemisphere.

Under what conditions will a hurricane lose its fuel supply?

What factors control the storm surge associated with a hurricane?

Lab 13

Climate Controls

Materials Needed

- world atlas

Introduction

What causes predictable patterns in temperature and precipitation? *Climate* is a statistical summary of atmospheric processes over an extended period of time. Viewed broadly, climate results from the physical interaction between the earth's atmosphere, water bodies, ice cover, and land surface. This lab examines seasonal temperature and precipitation patterns with respect to the physical controls causing them.

Climatic Data

We begin with definitions for some temperature and precipitation variables:

- *Average daily temperature.* The average of the day's minimum and maximum temperatures.
- *Average monthly temperature.* The average of the average daily temperatures in the month.
- *Average annual temperature.* The average of the 12 average monthly temperatures.
- *Annual temperature range.* The difference between the warmest average monthly temperature and the coolest average monthly temperature.
- *Monthly precipitation.* The total precipitation falling during the month.
- *Annual precipitation.* The total precipitation falling during the year.
- *Normal temperature.* Generally denoting the mean monthly or annual temperature over a 30-year period ending at the close of the last decade (e.g., 1971–2000).
- *Normal precipitation.* Generally denoting the mean monthly or annual precipitation over a 30-year period ending at the close of the last decade (e.g., 1971–2000).

	Jan	Feb	Mar	Apr	May	Jun	Jul	Aug	Sep	Oct	Nov	Dec
T (°C)	2.7	3.2	7.1	13.2	18.8	23.4	25.7	24.7	20.9	15.0	8.7	3.4
P (cm)	7.7	6.3	8.2	8.0	10.5	8.2	10.5	12.4	9.7	7.8	7.2	7.1

Summarizing Climatic Data

The table above shows normal monthly temperature and precipitation data for Washington, D.C. (latitude: 38° 50' N, longitude: 77° W).

The *climograph* below shows Washington's seasonal temperature and precipitation patterns. The line graph shows the annual march of temperature in degrees Celsius (left scale), and the bar graph shows monthly precipitation in centimeters (right scale).

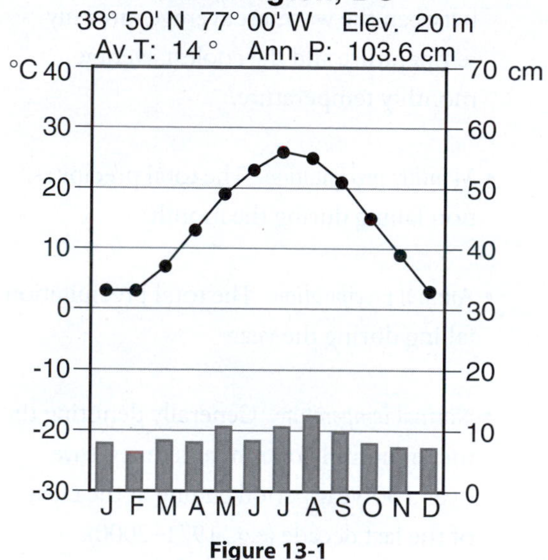

Figure 13-1

1. *Calculate the following statistics:*

Washington's average annual temperature: ______°C
Washington's annual temperature range: ______°C
Washington's annual precipitation: ______ *cm*

2. *Does Washington, D.C. have a marked wet season?*

3. *What other information can be gleaned from the climograph?*

Other Sample Climographs

4. *Figures 13-2 through 13-4 show three unidentified climographs. Indicate the pattern and range of temperature and precipitation for each. Is the station tropical, mid-latitude, or polar? Is there a marked rainy season?*

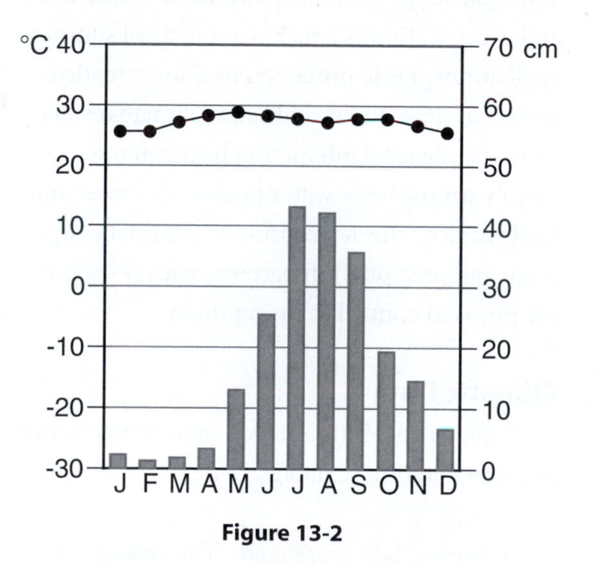

Figure 13-2

Temperature pattern and range:

Precipitation pattern and range:

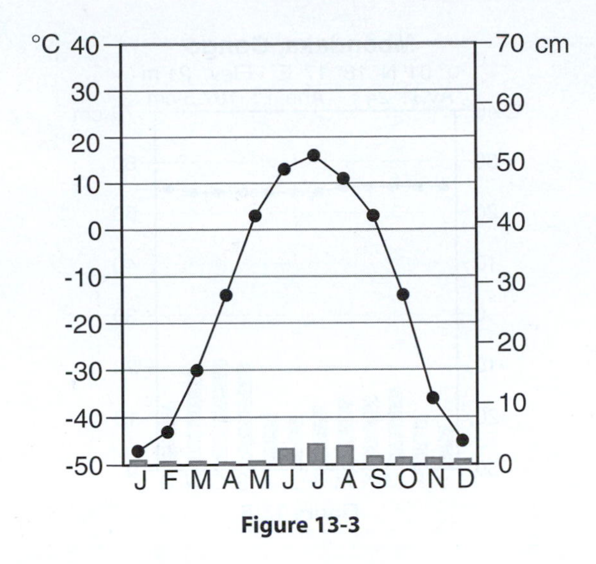

Figure 13-3

Temperature pattern and range:

Precipitation pattern and range:

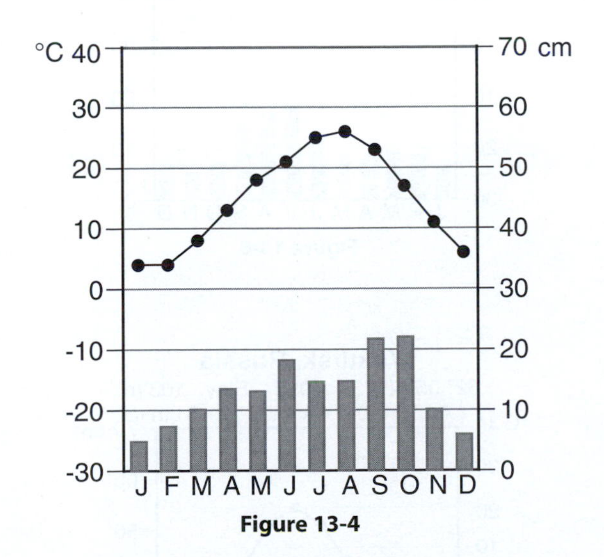

Figure 13-4

Temperature pattern and range:

Precipitation pattern and range:

Controls on Climate

We can generally attribute seasonal temperature and precipitation patterns to the influence of some combination of six climatic controls:

- Latitude
- Land and water
- Geographic position and prevailing winds
- Mountains and highlands
- Ocean currents
- Pressure and wind systems

In the rest of this lab we will examine each of these controls in turn.

Latitude

Latitude is the most important control on seasonal temperature variation. Figures 13-5 through 13-7 illustrate the differences in annual temperature range at tropical, mid-latitude, and polar sites, respectively.

5. *Why does the annual temperature range increase as the latitude increases?*

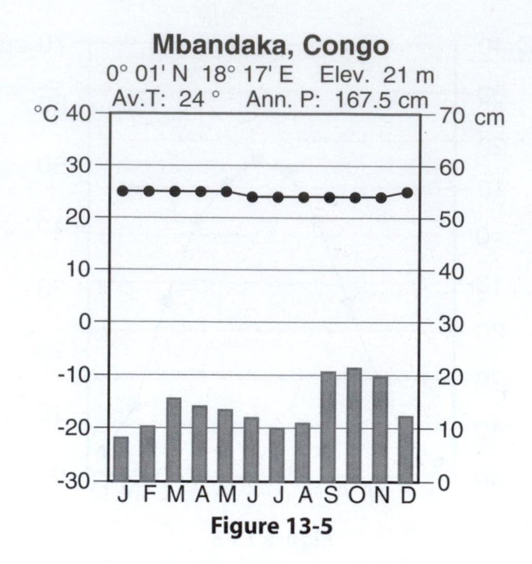

Figure 13-5

6. *Although this situation is not always true, the three climographs show decreasing precipitation with latitude. Why is there potential for greater precipitation at lower latitudes?*

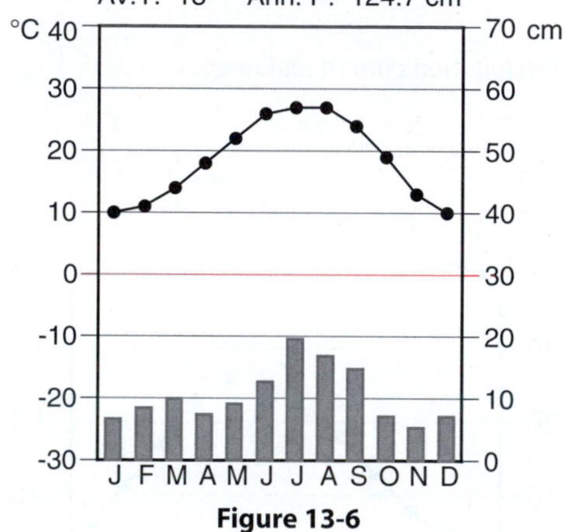

Figure 13-6

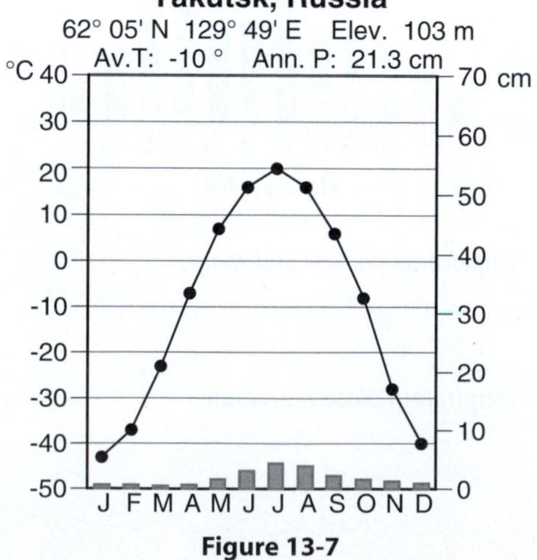

Figure 13-7

Land and Water

Land heats and cools more rapidly than water. Therefore, seasonal temperature variation is moderated at locations influenced by large water bodies. In these marine climates summers generally do not get excessively hot, and winters do not get excessively cold. By contrast, areas farther away from large water bodies experience greater seasonal temperature variation. Extremes are greater in these continental climates. Compare the marine climate of Reykjavik with the continental climate of Yakutsk.

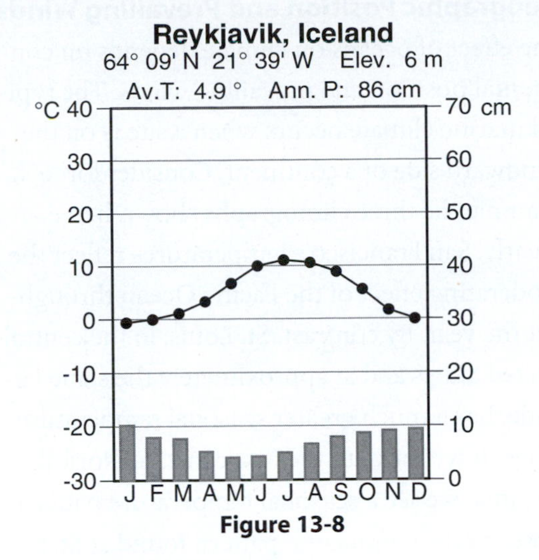

Figure 13-8

7. *What is the average annual temperature* ***range*** *at*
   Reykjavik? __________°C
   Yakutsk? __________°C

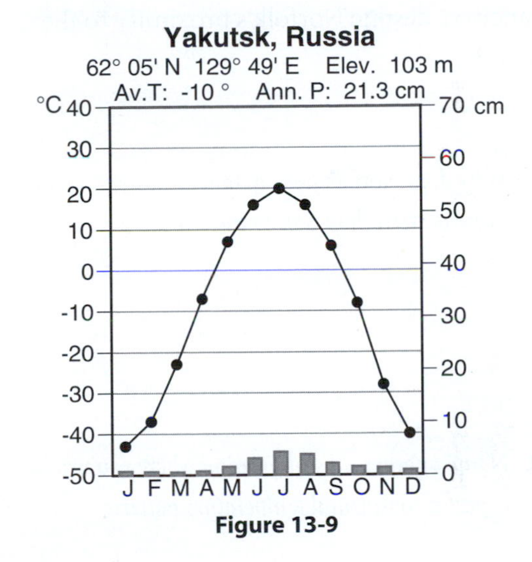

Figure 13-9

8. *Name two other places that would have different temperature patterns because of their contrasting proximity to the ocean. (Choose places with similar latitudes.)*

Geographic Position and Prevailing Winds

The effect of oceans on climate depends on continental position and prevailing winds. The typical marine climate occurs when a site is on the windward side of a continent. Consider for example the three climographs shown here. Clearly, San Francisco's temperatures reflect the moderating effect of the Pacific Ocean throughout the year. By contrast, St. Louis, in the central United States and at approximately the same latitude, has a much greater seasonal temperature range. If we move to the east coast, to Norfolk, Virginia, we see a seasonal temperature pattern closer to the continental pattern found at St. Louis than to the marine pattern at San Francisco, despite Norfolk's proximity to the Atlantic.

9. *Why does Norfolk have a more continental temperature pattern than San Francisco?*

10. *Name another coastal location where you would expect a continental temperature pattern.*

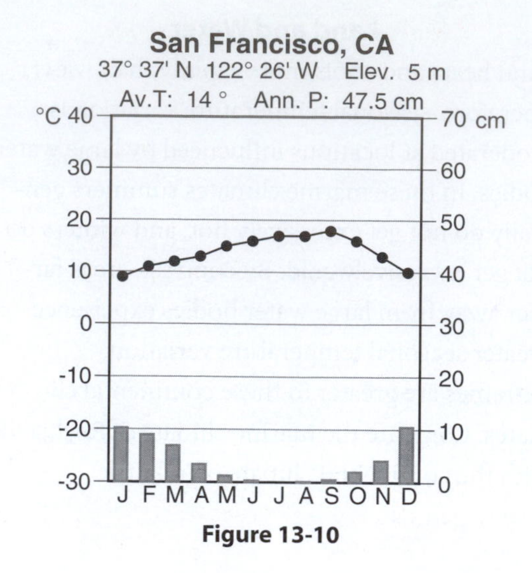

Figure 13-10

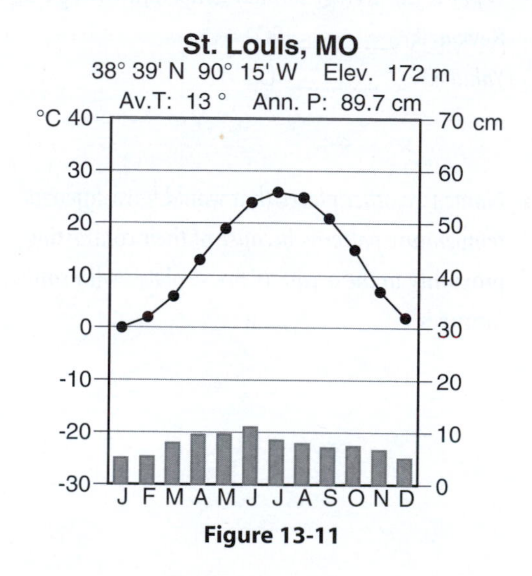

Figure 13-11

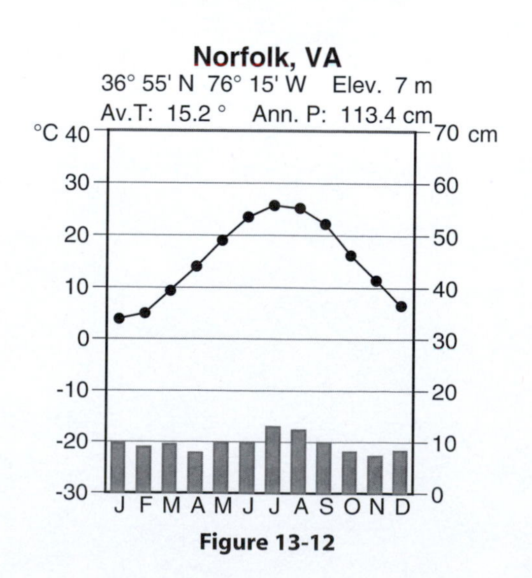

Figure 13-12

Mountains and Highlands

Mountains often modify airflow. In the mid-latitudes, where westerly winds predominate, mountain ranges that are oriented in a north-south direction affect the temperature and moisture characteristics of air flowing over them. Rising air on the *windward* flanks of a mountain range cools due to adiabatic expansion, producing clouds and precipitation on the windward side. As air descends on the *leeward* side of mountains it warms adiabatically, thus suppressing clouds and precipitation. The climatic contrast between Seattle and Spokane, Washington, illustrates the influence of the Cascade Range in the northwestern United States.

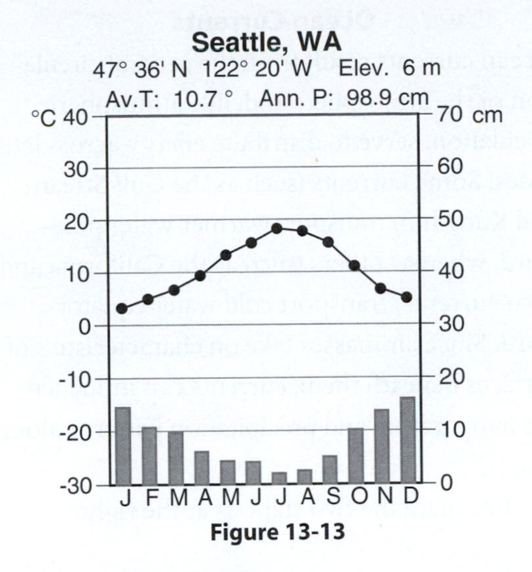

Figure 13-13

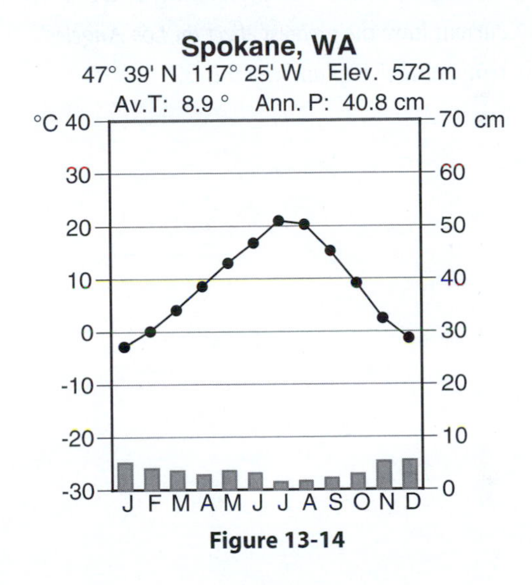

Figure 13-14

11. *Account for the difference in the annual temperature range between Seattle and Spokane.*

12. *Name two other cities that you suspect would have contrasting temperature and precipitation patterns because they are separated by a mountain range.*

Ocean Currents

Ocean currents result from the general circulation of the atmosphere and, like atmospheric circulation, serve to distribute energy across latitudes. Some currents (such as the Gulf Stream and Kuroshio) transport warmer water poleward, whereas others (such as the California and Peru currents) transport cold water equatorward. Since air masses take on characteristics of surfaces beneath them, currents can influence the temperature and precipitation patterns along coasts.

Compare the two stations at the right.

13. *During what season does proximity to a cold ocean current have the greatest effect on Los Angeles' temperature? Explain why this is.*

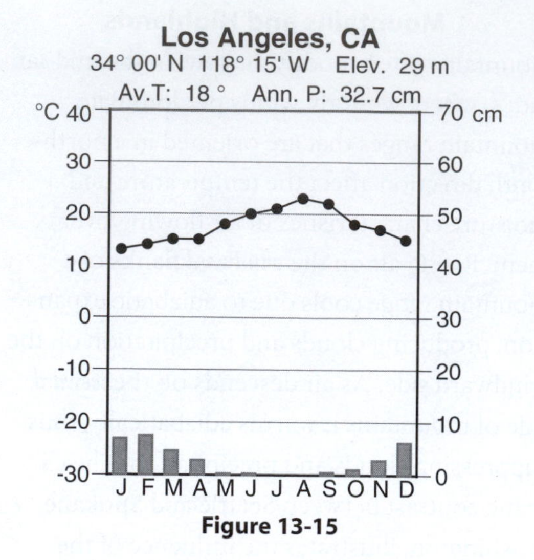

Figure 13-15

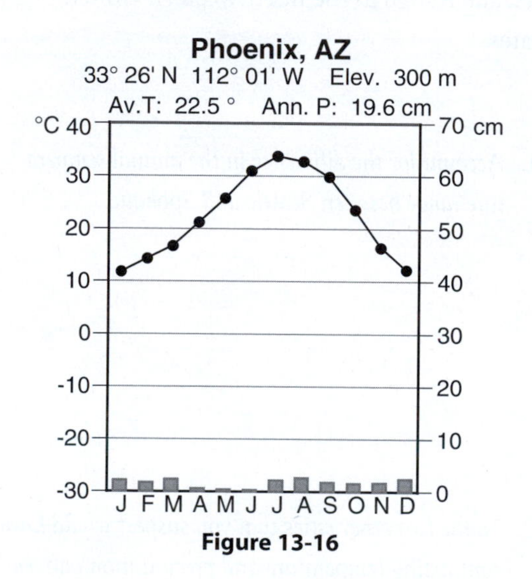

Figure 13-16

Now compare Charleston with Atlanta.

14. *During what season does proximity to a warm ocean current have the greatest effect on Charleston's temperature? Explain why this is.*

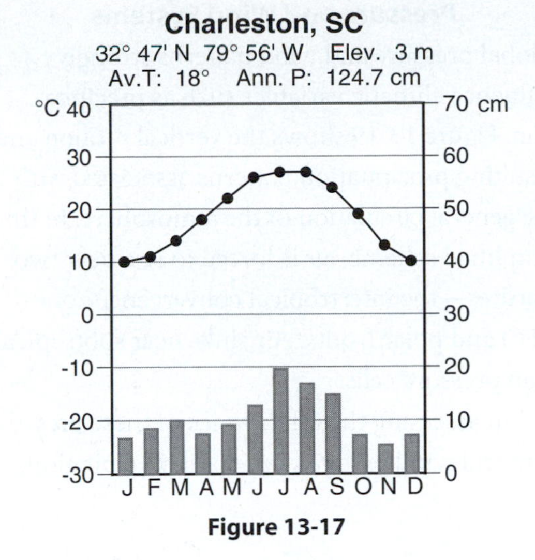

Figure 13-17

Sea-surface temperatures affect evaporation rates and atmospheric stability, since the lower atmosphere takes on temperature and moisture characteristics from its underlying surface.

15. *Compare the precipitation amounts between Los Angeles (Figure 13-15) and Charleston. Explain how the ocean currents affect evaporation rates and stability, contributing to the precipitation differences.*

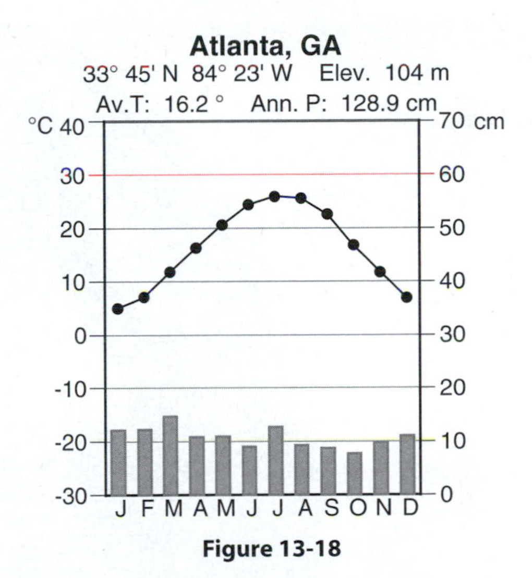

Figure 13-18

Ocean currents only partially explain the precipitation differences between Los Angeles and Charleston. We turn now to another very important climate control.

Pressure and Wind Systems

Global pressure and wind patterns strongly influence climatic variables such as precipitation. Figure 13-19 shows the vertical motion and resulting precipitation patterns associated with the general circulation of the atmosphere. In this simplified scheme, air is forced to rise near two features—the intertropical convergence zone (ITC) and polar fronts. Air sinks near subtropical high pressure cells.

In assessing the importance of these pressure and wind patterns on world precipitation, we must also consider their seasonal migration. Pressure and wind belts shift north and south through the year with the sun's direct rays, thus creating a seasonality to precipitation at many locations. Compare this idealized version the pressure and wind systems with the real-life patterns shown in Figure 8-18. Note the seasonal migration of the systems.

Now consider the precipitation patterns at Dakar, Senegal (Figure 13-20), and Rome, Italy (Figure 13-21): both locations have a well-defined rainy season.

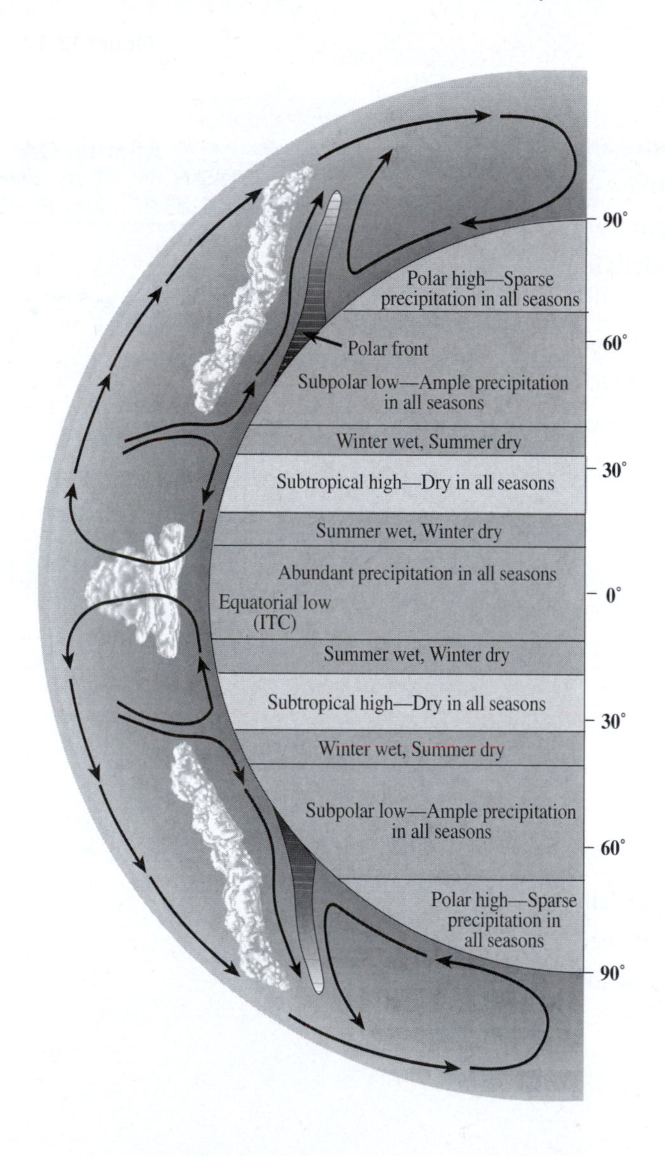

Figure 13-19. Schematic illustration of zonal precipitation patterns.

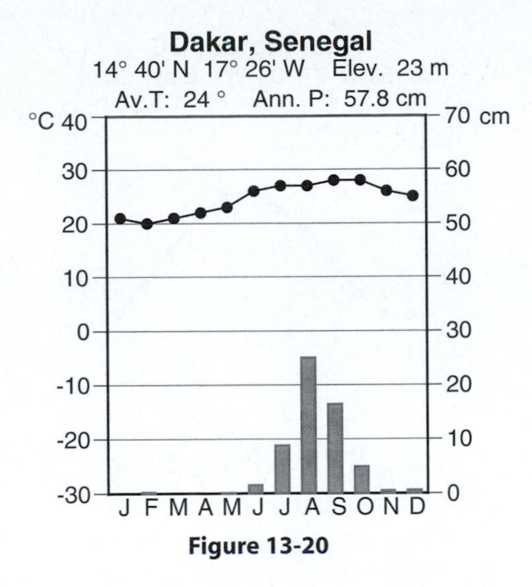

Figure 13-20

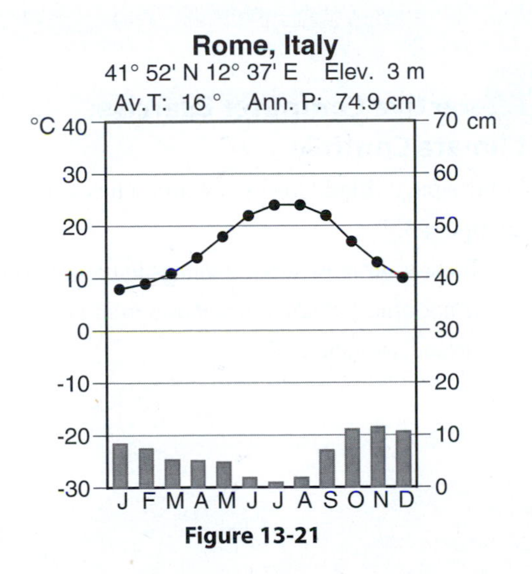

Figure 13-21

16. *What causes the rainy and the dry seasons in Dakar?*

17. *Explain why Rome has a rainy and a dry season.*

18. *Subsidence is typically weaker to the west and stronger to the east of the high pressure center. Explain how this fact, plus consideration for wind patterns and ocean temperatures, contributes to summer dryness in Los Angeles. By contrast, how do subtropical high pressure, winds, and ocean temperatures contribute to the very different summer precipitation pattern in Charleston?*

Controls on Washington, D.C.'s Climate

Four of the six climate controls help to explain Washington, D.C.'s temperature and precipitation patterns.

19. *Although Washington is close to the Atlantic Ocean, its seasonal temperature pattern reflects that of a more continental position. Why?*

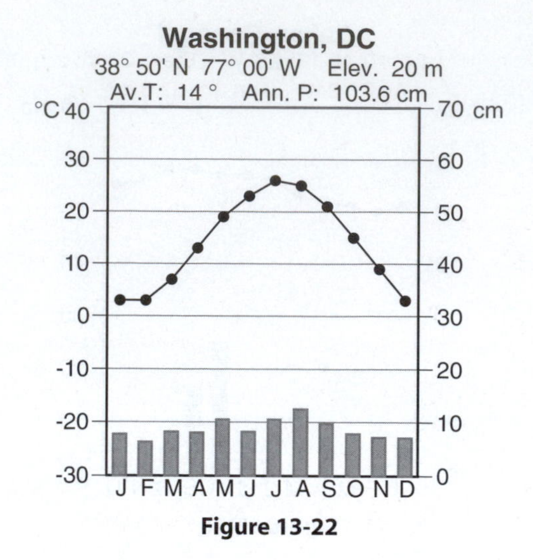

Figure 13-22

20. *Which ocean current affects Washington's climate?*

21. *During the summer months strong high pressure cells exist off North America's east and west coasts. How do these high pressure cells combine with the different ocean currents off the coasts to cause a summer dry season in San Francisco and plentiful rain in Washington, D.C.?*

Interactive Computer Exercise: Climate Controls

On the page titled Land vs. Water, click the Europe tab.

22. *What is the most striking change in the seasonal temperature pattern as one moves from Dublin, Ireland, to Samara, Russia?*

On the page titled Geographic Position, click the Comparisons tab.

23. *Describe the general differences in seasonal temperature and precipitation patterns between the east and west coasts of North America.*

On the page titled Mountains, click the Daily Temperatures tab:

24. *The Denver and Evergreen comparison illustrates how elevation affects minimum temperature more than maximum temperature. How might differences in the upward and downward fluxes of longwave radiation help to explain this phenomenon?*

On the page titled Questions : North America:

25. *How do temperature and precipitation patterns vary as one moves from New Orleans north to the interior of Canada? Explain what causes the differences.*

On the page titled Questions : NA West Coast:

26. *Santiago and San Diego receive about the same amount of precipitation annually, but how does the seasonal timing of precipitation differ? Why does it differ?*

On the page titled Questions : Africa:

27. *Why is it so dry in Windhoek, Namibia?*

28. *Why is it so wet in Kisangani, Democratic Republic of Congo?*

On the page titled Questions : World:

29. *Choose any two cities from the world map and explain their annual temperature and precipitation patterns with respect to any of the climate controls discussed in the computer module.*

Interactive Computer Exercise: Africa ITCZ

Explore this module by clicking or rolling over stations on the map and by clicking individual months or the **Play** button.

30. *Kisangani, Democratic Republic of Congo, has two precipitation maxima during the year. In what months do these occur? Explain why they occur then.*

31. *Notice the similarity between the climographs of Bambari, Central African Republic, and Kananga, Democratic Republic of Congo. During which season does most precipitation fall at these two sites?*

32. *Why is it so much wetter in Bambari than it is in Abéché, Chad? What accounts for the annual temperature pattern at Abéché?*

Review Questions

Some tropical and some mid-latitude locations experience marked rainy seasons. How do the causes of these rainy seasons generally differ between the tropics and the mid-latitudes?

Today we can find glaciers in the tropics. How is this possible?

Although we've examined each climate control separately, some of the cities chosen for this lab illustrate how controls can be closely related. Use the temperature and precipitation patterns for Charleston and San Francisco to show how land and water, geographic position, and ocean currents are interrelated controls.

Lab 14

Climate Classification

Materials Needed

- world atlas
- climate data

Introduction

How can we organize climates into meaningful categories? Wladimir Köppen devised a climate classification scheme based on monthly temperature and precipitation—the main determinants of plant range. It is one of many schemes designed to organize the world's climates into distinct groups. This lab examines Köppen's major climate groups with respect to the physical controls on each. The climate types include:

A Climates—Tropical
- The Wet Tropics (Af)
- Tropical Monsoon (Am)
- Tropical Wet and Dry (Aw)

B Climates—Dry
- Desert (BW)
- Steppe (BS)

C Climates—Humid Mesothermal
- Humid Subtropical (Cfa)
- Marine West Coast (Cfb)
- Mediterranean (Cs)

D Climates—Humid Microthermal
- Humid Continental (Dfa, Dfb, Dwa, Dwb)
- Subarctic (Dfc, Dfd, Dwc, Dwd)

E Climates—Polar
- Tundra (ET)
- Ice Cap (EF)

A Climates—Tropical

- Warm all year (typical annual average: 25°C; coldest month > 18°C)
- Small annual temperature range
- High annual precipitation
- Rainfall more variable than air temperature

Af—Tropical Wet (Figure 14-1)

- No dry season (precipitation exceeds 6 cm every month)

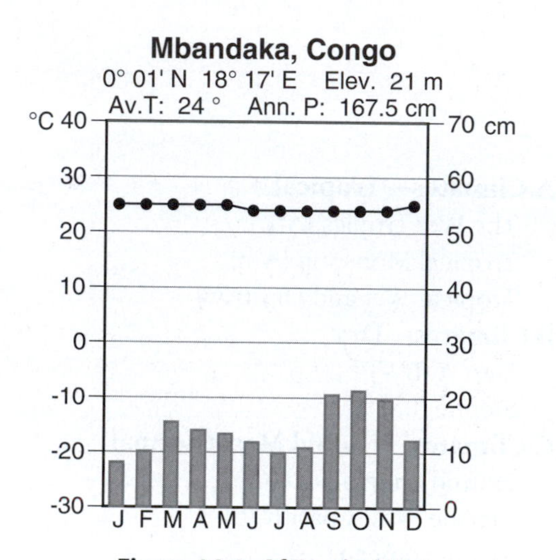

Figure 14-1. Af Tropical Wet.

1. *How does Mbandaka's latitude account for its very small annual temperature range?*

2. *What accounts for the double precipitation peak in Mbandaka (March–April and September–November)?*

Am—Tropical Monsoon

- Very short dry season
- Less than 6 cm precipitation in driest month, but high annual total

Aw—Tropical Wet and Dry (Figure 14-2)

- Warmest months occur just before wet season
- Well-defined winter dry season (w = winter dry season)
- Less than 6 cm precipitation in driest month

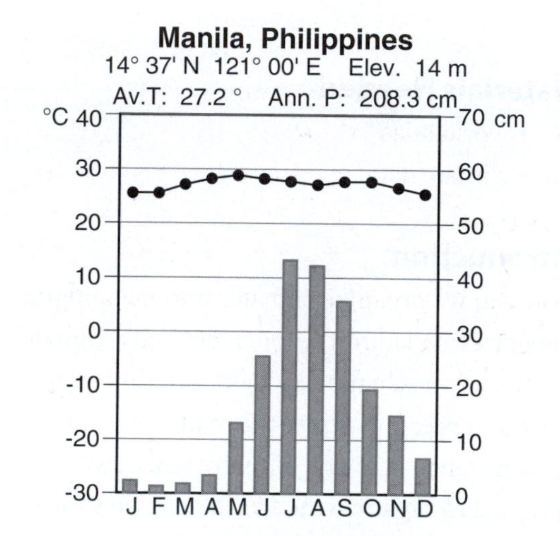

Figure 14-2. Aw Tropical Wet and Dry.

3. *Explain the seasonal precipitation pattern at Manila in terms of pressure and wind systems. What causes the summer rains? What causes the winter dry season?*

4. *Why do you think May is slightly warmer than June, July, or August?*

B Climates—Dry

- Moisture deficit: potential evaporation exceeds precipitation
- Unreliable year-to-year precipitation

BW—Desert (Figure 14-3)

- True desert; arid
- BWh—Subtropical or "hot" deserts (average annual temperature > 18°C)
- BWk—Mid-latitude deserts (average annual temperature < 18°C)

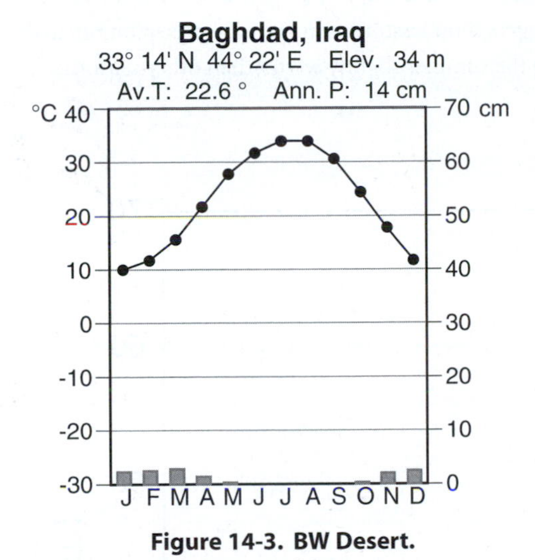

Figure 14-3. BW Desert.

5. *How does Baghdad's latitude influence its annual temperature range?*

6. *Is Baghdad a subtropical or a mid-latitude desert?*

7. *What is the dominant pressure system influencing Baghdad's annual precipitation pattern?*

8. *What storm systems produce a little rain during the winter months?*

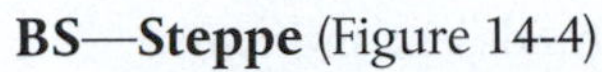

BS—Steppe (Figure 14-4)

- Semi-arid transition zone between arid and humid climates
- BSh and BSk defined as above

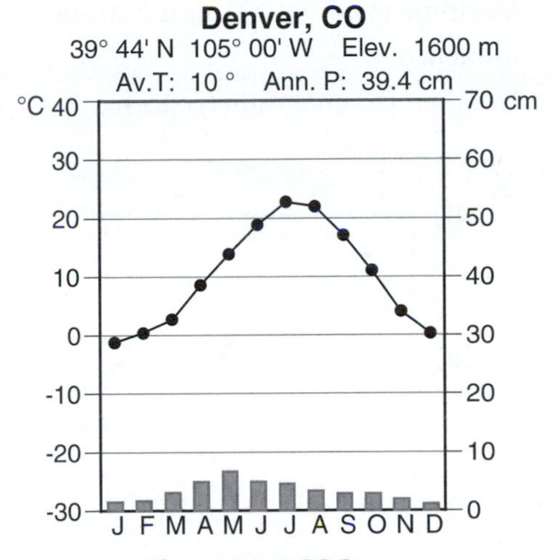

Figure 14-4. BS Steppe.

9. *What geographic factors cause Denver's climate to be relatively dry?*

C Climates—Humid Mesothermal

- Moderate seasonal temperature range
- Relatively mild winters (Coolest month is below 18°C but above -3°C)

Cfa—Humid Subtropical

- Found on eastern side of continents, under influence of western side of subtropical highs
- Hot, sultry summers (warmest month over 22°C)
- Ample annual precipitation with no marked dry season
- Maritime tropical (mT) air is a strong influence
- Frequent summer thunderstorms

10. *Use the following brief climatic description to sketch a climograph for a humid subtropical site.*

The mid-latitude location (35° N) results in a moderate seasonal temperature range. It is located along the east coast of a continent and is influenced by a warm ocean current. The site is far away from any significant mountain ranges. Precipitation is ample throughout the year, with no marked dry season: the site receives an average of 125 cm of precipitation annually. During the winter months, mid-latitude cyclones and their associated fronts cause regular rainfall at the site. During warmer months, an oceanic high pressure center to the east brings maritime tropical air to the site. This unstable air triggers thunderstorms and intense precipitation, making the summer slightly wetter than other seasons.

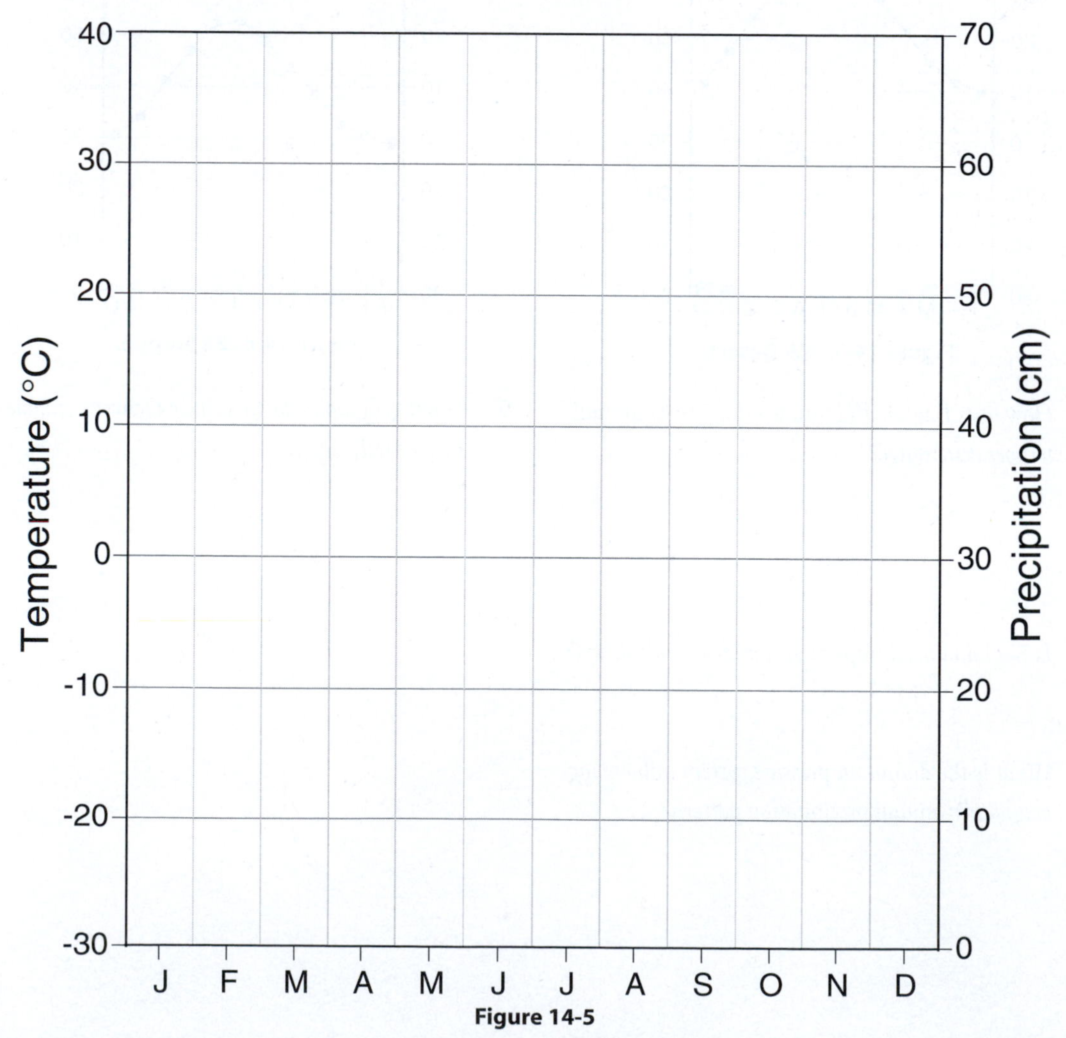

Figure 14-5

Cfb—Marine West Coast (Figure 14-6)

- Mild winters
- Cool summers (at least 4 months above 10°C, warmest month below 22°C)
- Ample annual precipitation; slightly reduced during summer months
- Maritime Polar (mP) air masses dominant

Cs—Mediterranean

(Dry-Summer Subtropical; Figure 14-7)

- Equatorward of marine west coast climate and poleward of subtropical steppe climate
- Well-defined summer dry season (s = summer dry season)
- Mild winters, warm (Csb) or hot (Csa) summers

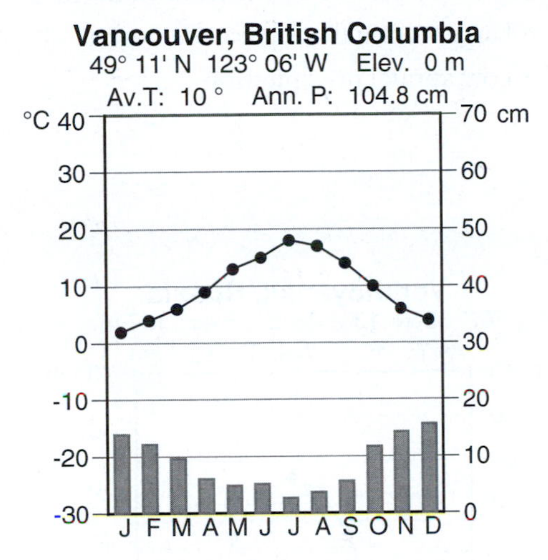

Figure 14-6. Cfb Marine West Coast.

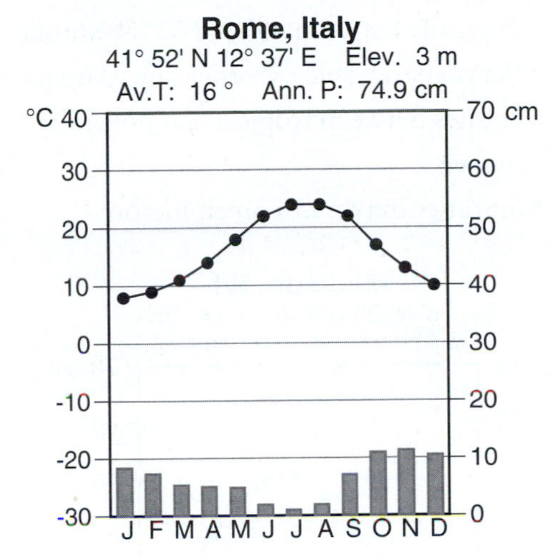

Figure 14-7. Cs Mediterranean.

11. *How does the annual temperature range at Vancouver compare with that at most other locations at a similar latitude? Why does it differ?*

12. *How does the seasonal migration of pressure and wind systems cause reduced summer precipitation at Vancouver?*

13. *What pressure patterns and wind systems cause Rome's precipitation to increase during the winter months?*

14. *What pressure patterns and wind systems suppress summer precipitation in Rome?*

D Climates—Humid Microthermal

- Poleward and interior of C climates
- Continental climates found in the Northern Hemisphere only
- Large seasonal temperature range
- Severe winters (coldest month below -3°C, warmest above 10°C)

Dfa, Dfb, Dwa, Dwb—Humid Continental (Figure 14-8)

- Typically between 40° and 55° N latitude
- Very "changeable" weather due to frequent clashes between tropical and polar air masses
- Summer maximum precipitation

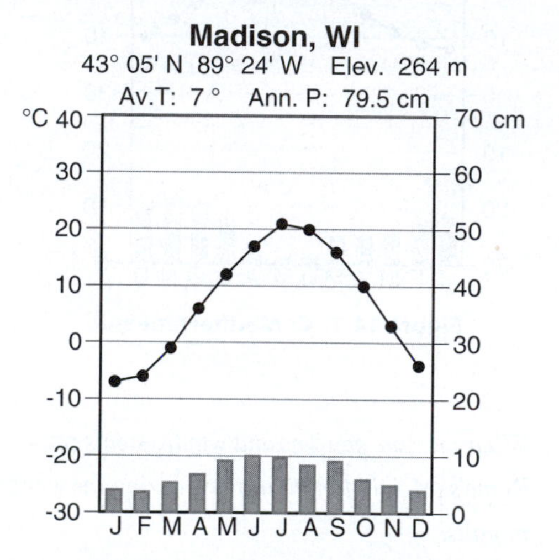

Figure 14-8. Humid Continental.

15. *What is the chief atmospheric moisture source for Madison and most of the central United States?*

16. *What wind direction would support the summer precipitation maximum in Madison and most central U.S. locations?*

17. *Why is this climate type rare in the Southern Hemisphere?*

Dfc, Dfd, Dwc, Dwd—Subarctic (Figure 14-9)

- North of humid continental climates
- Largest annual temperature range
- Low annual precipitation

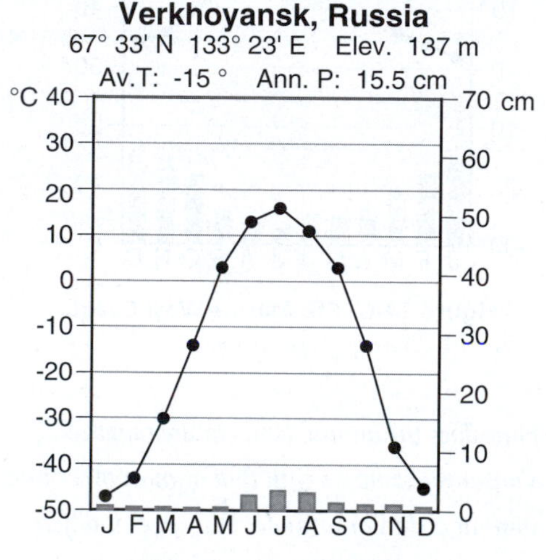

Figure 14-9. Subarctic.

18. *How do latitude and continental position cause such a large temperature range in Verkhoyansk?*

E Climates—Polar

- Warmest month < 10° C

ET—Tundra (Figure 14-10)

- Warmest month > 0°C
- Mostly Northern Hemisphere
- Precipitation typically < 25 cm

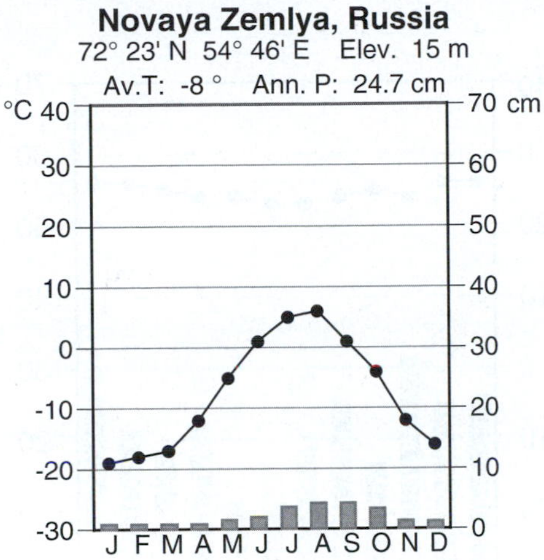

EF—Ice Cap (Figure 14-11)

- Warmest month < 0°C
- Precipitation typically 7.5–10 cm

Novaya Zemlya, Russia
72° 23' N 54° 46' E Elev. 15 m
Av.T: -8 ° Ann. P: 24.7 cm

°C 40 | 30 | 20 | 10 | 0 | -10 | -20 | -30
70 cm | 60 | 50 | 40 | 30 | 20 | 10 | 0
J F M A M J J A S O N D

Figure 14-10. ET Tundra.

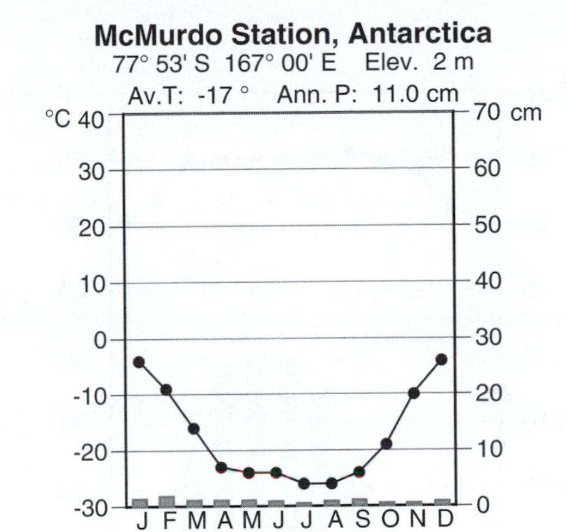

Figure 14-11. EF Ice Cap.

19. *From the perspective of vegetation, what is the significance of mean temperatures that exceed 0°C for a month or longer?*

20. *Since sun angle at McMurdo Station is never higher than 35½° above the horizon, solar intensity is low even in summer months. How might surface conditions at McMurdo station also prevent temperatures from getting very high?*

Identifying Climographs

21. *Match the following cities to the climographs below:*

    a. Iquitos, Peru—3° 39' S, 73° 18' W, elevation: 115 m
    b. Yuma, USA—32° 40' N, 114° 40' W, elevation: 62 m
    c. Calcutta, India—22° 32' N, 88° 22' E, elevation: 6 m
    d. Miami, USA—25° 45' N, 80° 11' W, elevation: 2 m

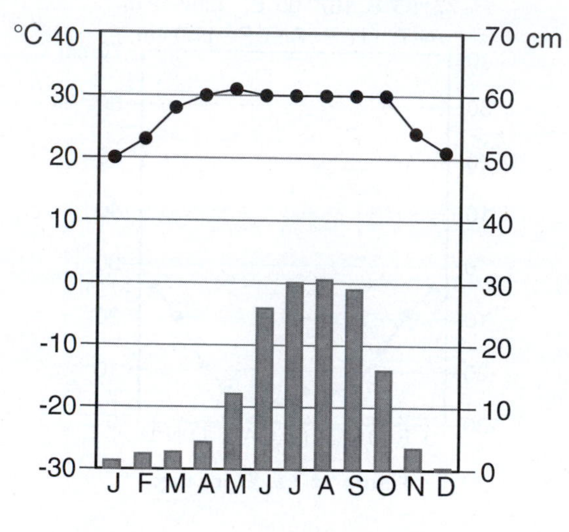

City ______________________

Climate type ______________

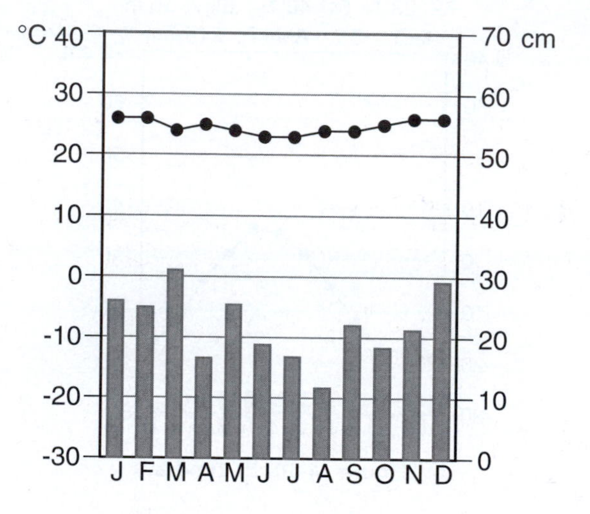

City ______________________

Climate type ______________

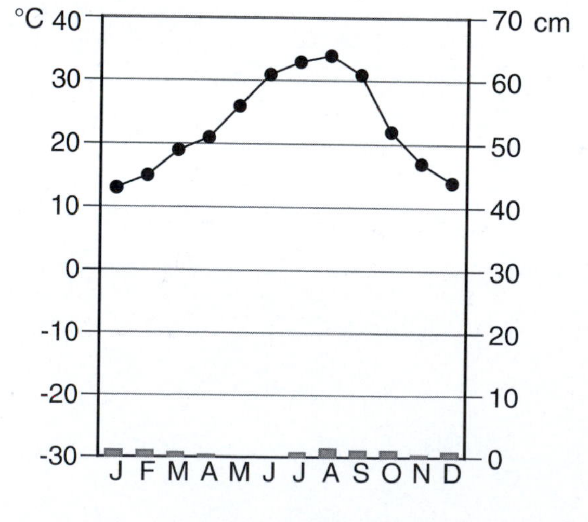

City ______________________

Climate type ______________

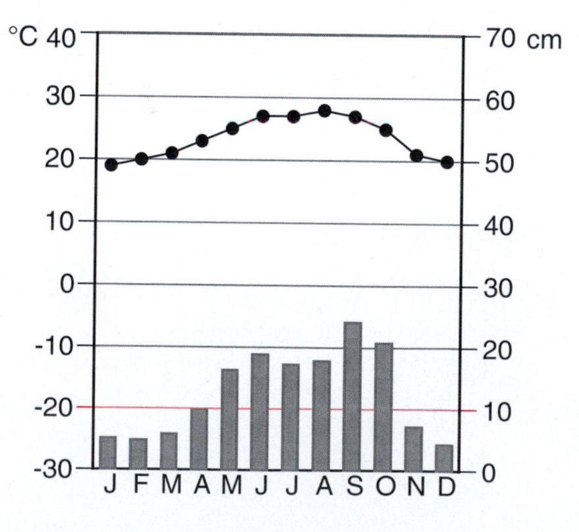

City ______________________

Climate type ______________

22. *Match the following cities to the climographs below:*

e. Singapore, Singapore—1° 22' N, 103° 52' E, elevation: 6 m

f. Lima, Peru—12° 06' S, 76° 55' W, elevation: 120 m

g. Sydney, Australia—33° 52' S, 151° 17' E, elevation: 42 m

h. Perth, Australia—31° 50' S, 116° 10' E, elevation: 60 m

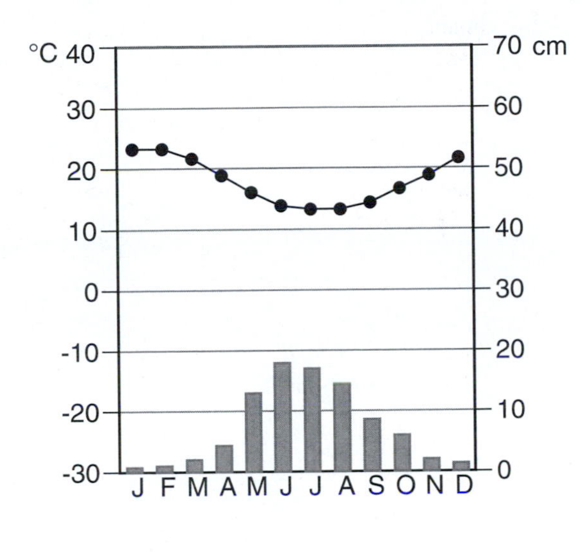

City ____________________

Climate type ____________________

City ____________________

Climate type ____________________

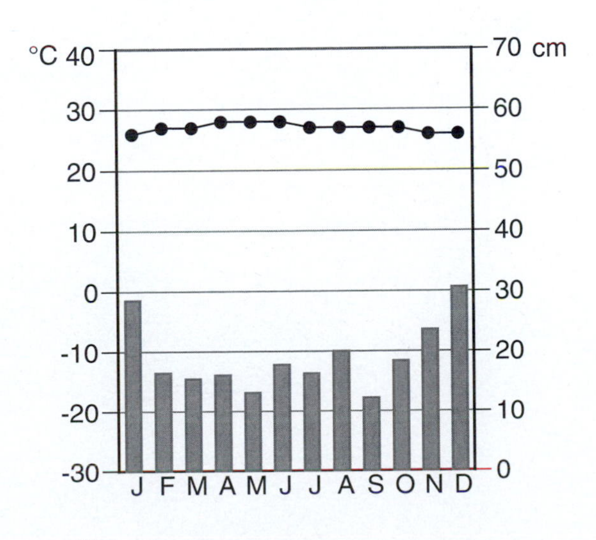

City ____________________

Climate type ____________________

City ____________________

Climate type ____________________

Making Your Own Climograph

23. *Describe the climate of any location in the world.*
    a. *You should begin by choosing a place and acquiring monthly temperature and precipitation data. This can be done using your textbook, climatic data references, the Internet, atlases, climate atlases, maps, or other data sources.*

    b. *On the opposite page, sketch a climograph for the city you chose.*

    c. *Describe the average temperature and precipitation patterns, and explain the location's climate with regard to climatic controls.*

    d. *Where information is available, identify and describe years that differ significantly from average conditions. You may also consider describing how vegetation patterns at your location reflect climate.*

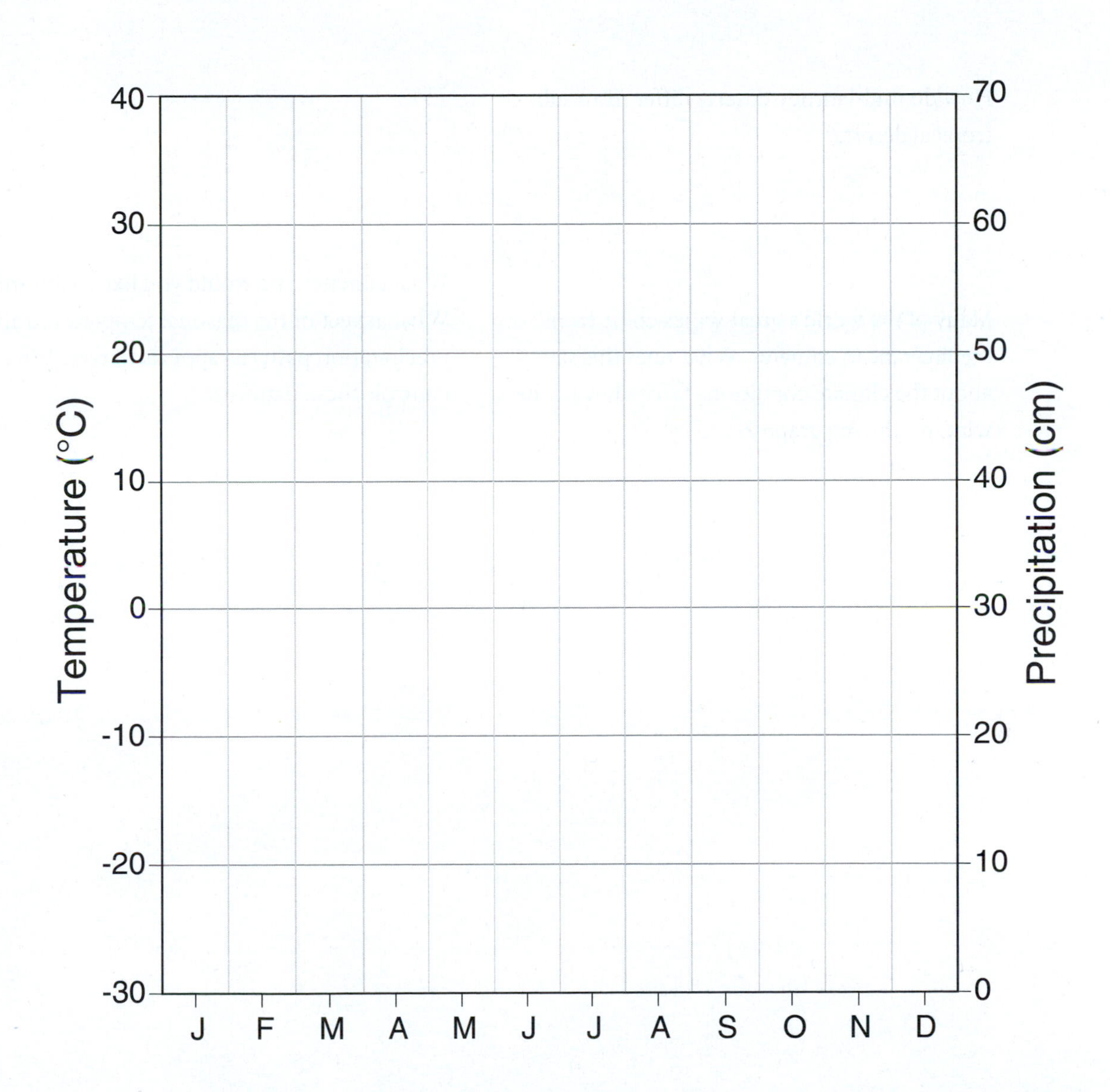
40
30
20
10
0
-10
-20
-30
Temperature (°C)
70
60
50
40
30
20
10
0
Precipitation (cm)
J
F
M
A
M
J
J
A
S
O
N
D

Review Questions

What aspects of climate differ between regions with rain forests and regions with savanna vegetation?

How do mid-latitude deserts differ from subtropical deserts?

Many of the world's great wines come from Mediterranean climates. What does this say about the climate conditions favored by the best wine-producing grapes?

The subarctic climate supports coniferous forests. How can a forest survive the cold temperatures and low precipitation that characterize this climate type?

What climate type would you like to live in? What aspect of the seasonal temperature and precipitation patterns appeals to you? What controls these patterns?

Lab 15

Climate Variability and Change

Introduction

Is our climate changing? How can we tell? With recent interest in global warming, climatic variability and change have become popular topics. This interest has, however, highlighted the uncertainty involved in our understanding of the climate. It can be difficult to identify climate variability and change and to determine their specific causes. Moreover, the very way we collect data may influence the historic climate record. Added to these difficulties, climate change often involves times scales extending beyond our written climate records, requiring scientists to develop proxy measures for past climates. This lab will focus on some of the methods climatologists use to detect climate changes and some of the problems they have with their data.

Hypothetical Temperature Curves

With statistics, we can examine a time series—i.e., how a particular climate variable changes over time. One way of examining a time series is to break out its component parts. Statistical analysis can reveal different aspects of the time series. There may be a regular *cycle* present (Figure 15-1), an *upward trend* (Figure 15-2), or *sudden changes* (Figure 15-3). The goal of statistical analysis is to uncover a link between these components and the factors that may influence or force them, as the forcing factors themselves may exhibit similar cycles, trends, and sudden changes. The process is complicated because there usually is a *random* component (Figure 15-4) to each time series.

1. *Name and explain any mechanisms that could produce cyclical climate fluctuations, a linear trend, or a sudden increase or decrease in temperature.*

2. *Figure 15-5 shows a sample temperature time series. Rather than analyzing this time series statistically, see if you can visually detect any of the components discussed previously. Label those you find.*

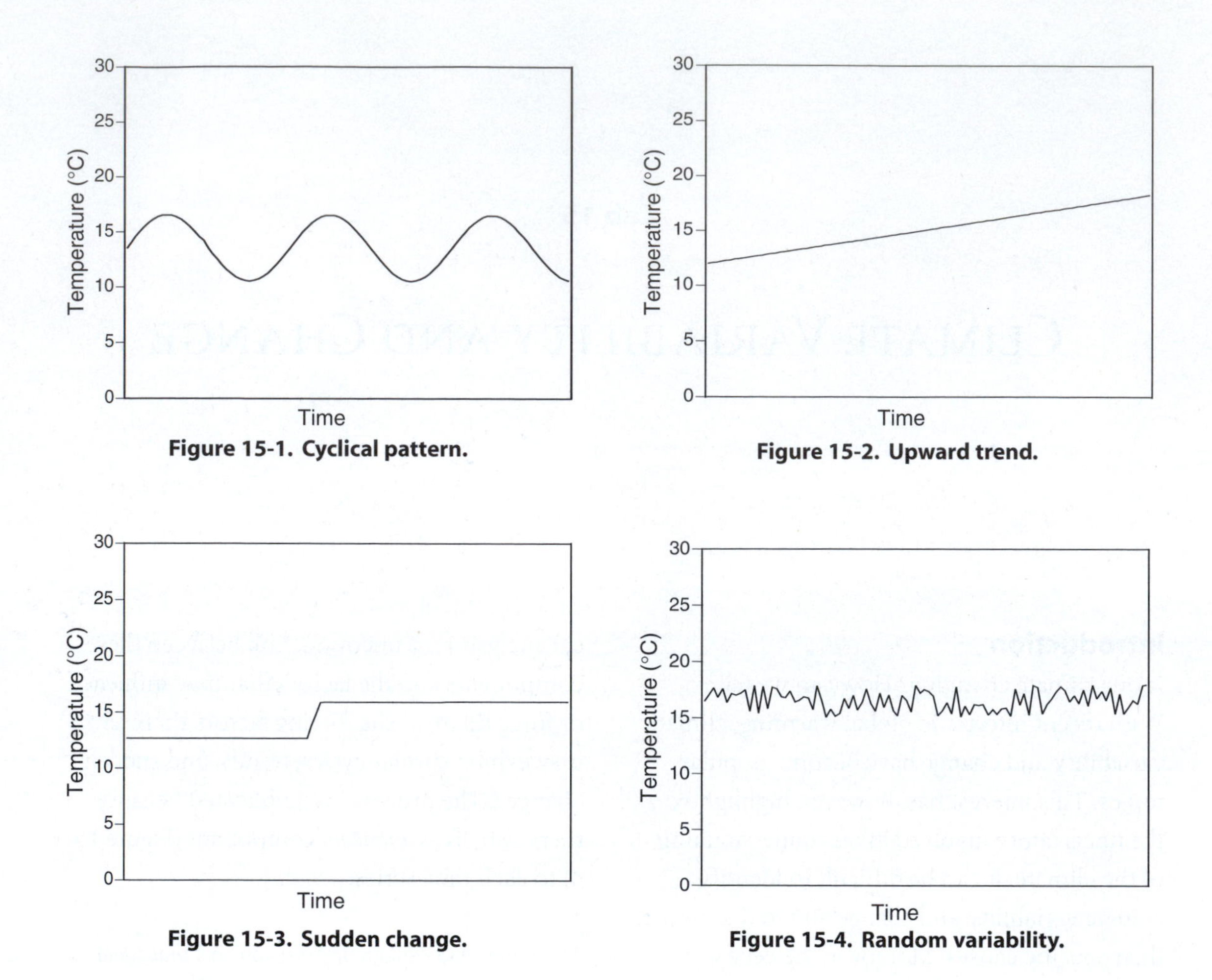

Figure 15-1. Cyclical pattern.

Figure 15-2. Upward trend.

Figure 15-3. Sudden change.

Figure 15-4. Random variability.

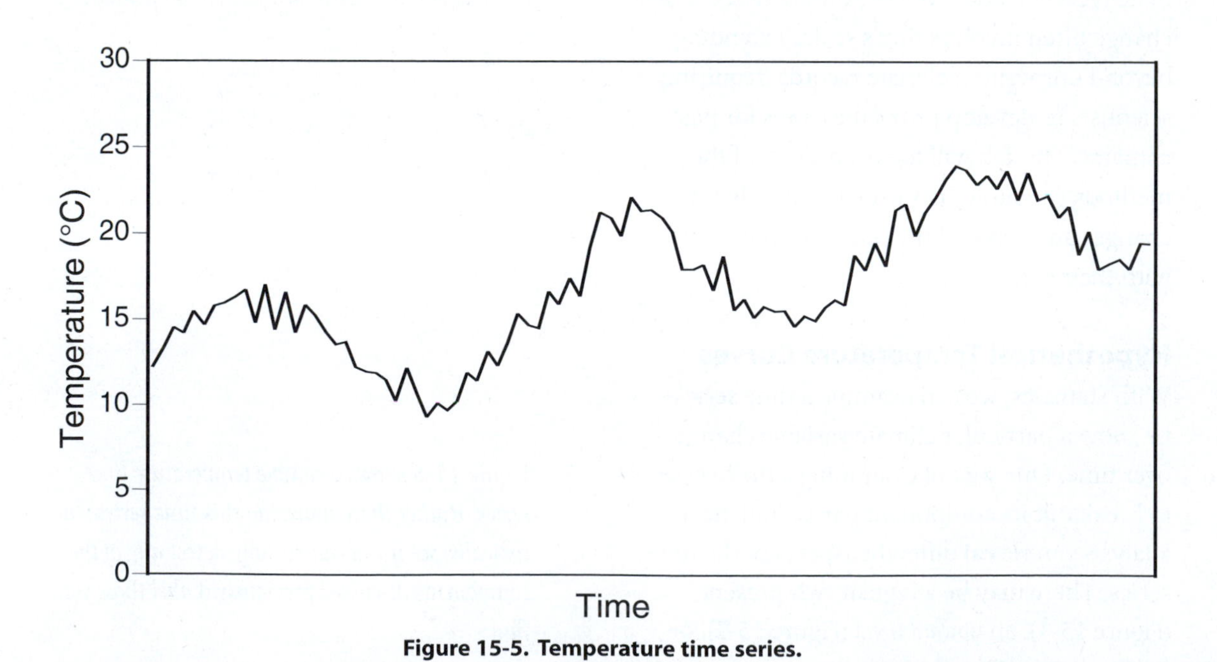

Figure 15-5. Temperature time series.

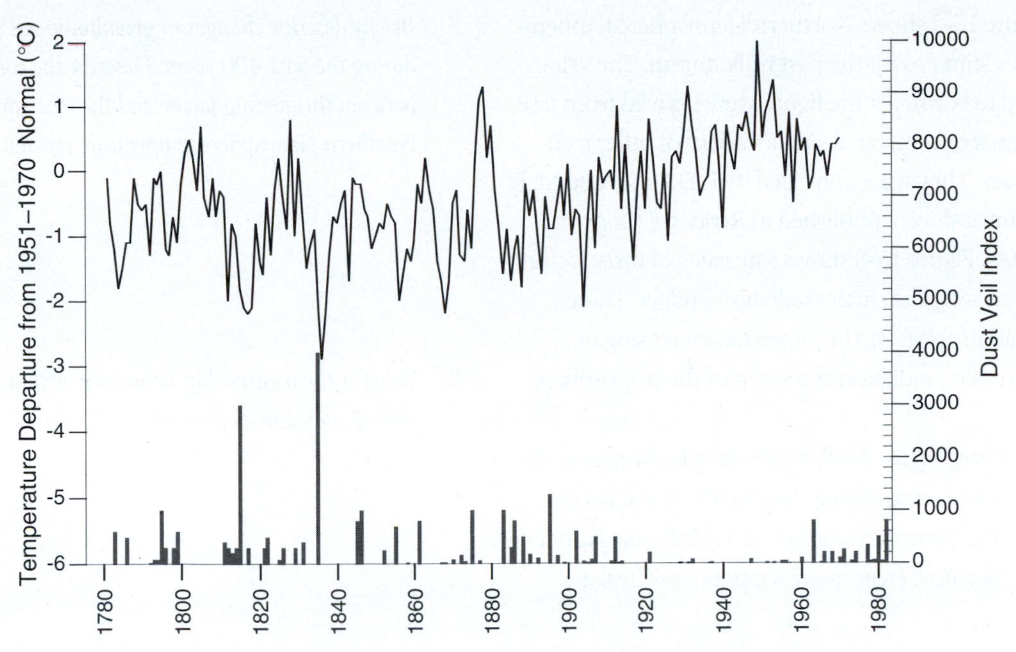

Figure 15-6. New Haven, CT, annual temperature variability.

3. *Examine the New Haven, Connecticut, annual temperature anomaly curve (Figure 15-6). The data are plotted as °C deviations from the 1951–1970 mean. Negative values show below-normal temperatures, positive values above-normal temperatures. Comment on any patterns in the data such as cycles, gradual trends, or dramatic changes.*

4. *Which century was generally warmer, the 19th or the 20th?*

5. *The bars in Figure 15-6 show the annual dust veil index, a proxy for volcanic activity. The two largest spikes correspond to the eruptions of Mts. Tambora, Indonesia (1815), and Coseguina, Nicaragua (1835). For how long and by how much does the New Haven temperature record respond to those events?*

Figure 15-7 shows Northern Hemisphere temperature estimates for the past millennium. The values used to construct the figure were derived from tree rings, ice cores, coral, and a variety of other techniques. They were compiled by P. D. Jones and M. E. Mann and were published in *Reviews of Geophysics* (2004). Figure 15-8 shows estimates of three factors influencing the earth's radiation budget. These *radiative forcings* can be viewed as increasing or decreasing radiation at the top of the troposphere.

6. *Using Figure 15-8, briefly describe changes in the solar forcing during the past 400 years. Do you detect a relationship between this forcing factor and Northern Hemisphere temperatures? Explain.*

7. *Briefly describe changes in greenhouse gas forcing during the past 400 years. Describe the correlation between this forcing factor and the reconstructed Northern Hemisphere temperature estimates.*

8. *What is the relationship between greenhouse gas forcing and aerosol forcing?*

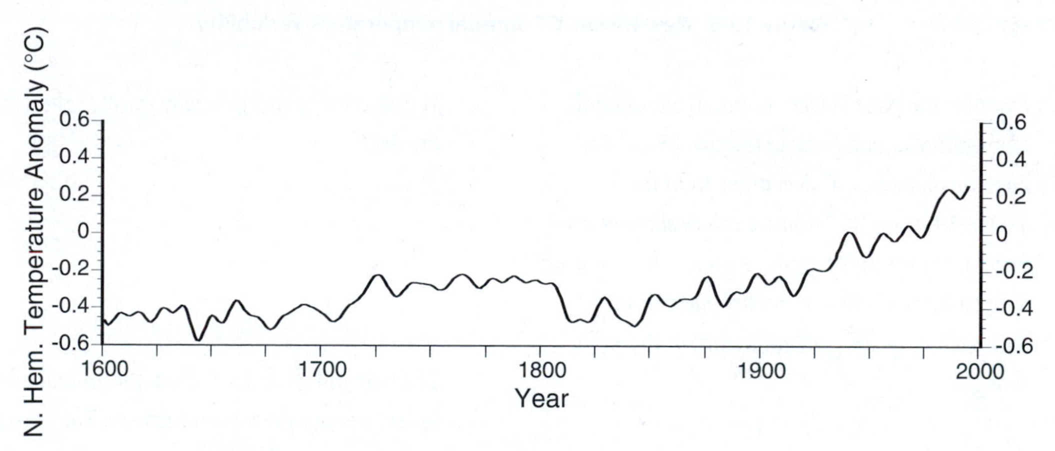

Figure 15-7. Northern Hemisphere temperature estimates.

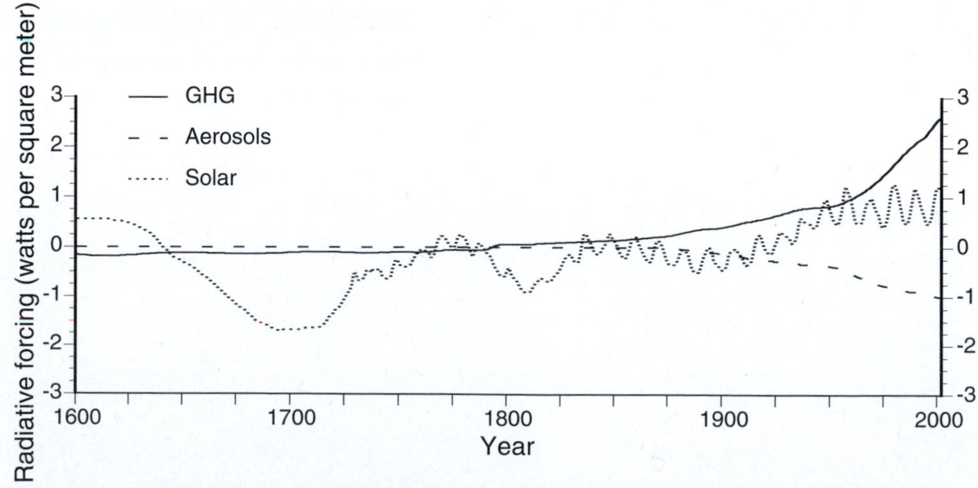

Figure 15-8. Radiative forcing estimates for greenhouse gases, aerosols, and solar radiation.

9. *Figure 15-9 shows the Northern Hemisphere instrumental temperature from 1856 to 2000 (solid line) against that predicted by a computer model that considers radiative forcing estimates. In what ways does the computer model output match the instrumental record? In what ways does it differ?*

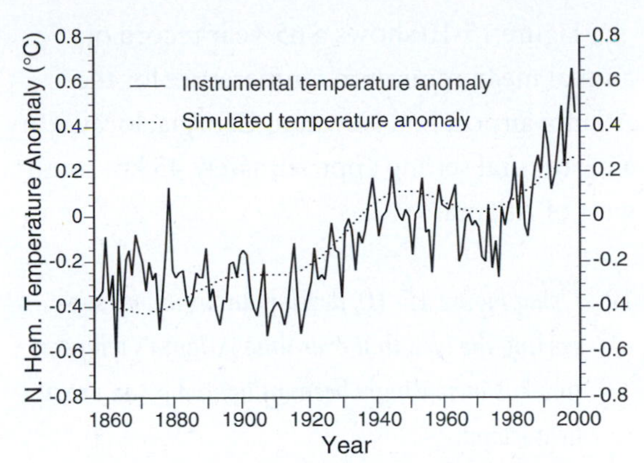

Figure 15-9. Simulated vs. instrumental temperature.

The Effect of Urbanization

Estimates of climatic variability and change during the past 150 years are primarily derived from the instrumental record (i.e., compiled from direct observations with instruments). This record is more precise than indirect evidence, but it is not without problems. Sites in Europe and North America are disproportionately represented in the early part of the record and perhaps skew what we know about global climate. The growth of cities during this period also creates problems with the record, since urban climates differ significantly from surrounding rural areas (Table 15-1).

10. *Why would average temperature generally be higher in the city than in surrounding rural areas?*

11. *Table 15-1 shows that regional-scale urban winds are generally lower than those in surrounding rural areas. What do you think causes this? By contrast, the microscale winds around buildings can be relatively fast. Why? How could low wind speeds contribute to the urban heat island?*

Table 15-1. Sample differences between urban and rural environments.

Variable	Urban Environment Compared with Rural Environment
Temperature	0.5°–1.5°C higher
Solar radiation	15%–30% less
Precipitation	5%–15% more
Wind speed	25% lower

Figure 15-10 shows a 65-year record of annual mean minimum temperature for the Atlanta airport and Newnan, Georgia, located in a more rural setting approximately 45 km southwest of Atlanta.

12. *Using Figure 15-10, develop an argument supporting the idea that over time Atlanta's temperature has increasingly been influenced by an urban heat island.*

13. *How could urban heat islands influence our understanding of temperature changes during the past 150 years?*

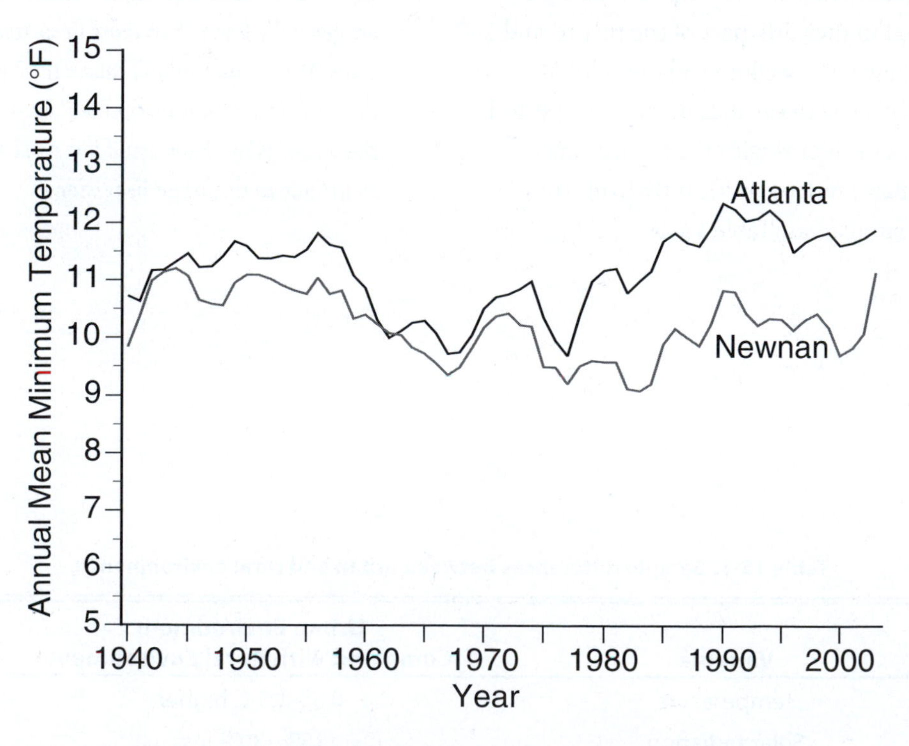

Figure 15-10. Atlanta and Newnan minimum temperatures, 3-year running mean, 1940–2004.

Data from a study in Orlando, Florida, show how a city can affect diurnal temperatures. Figure 15-11 below compares the January 11 and 12 hourly temperatures at stations in Orlando's urban core with those from park locations on the outskirts of the metropolitan area.

14. *During what hours is the urban heat island most prominent? Why do you think it is more clearly defined during certain times of the day?*

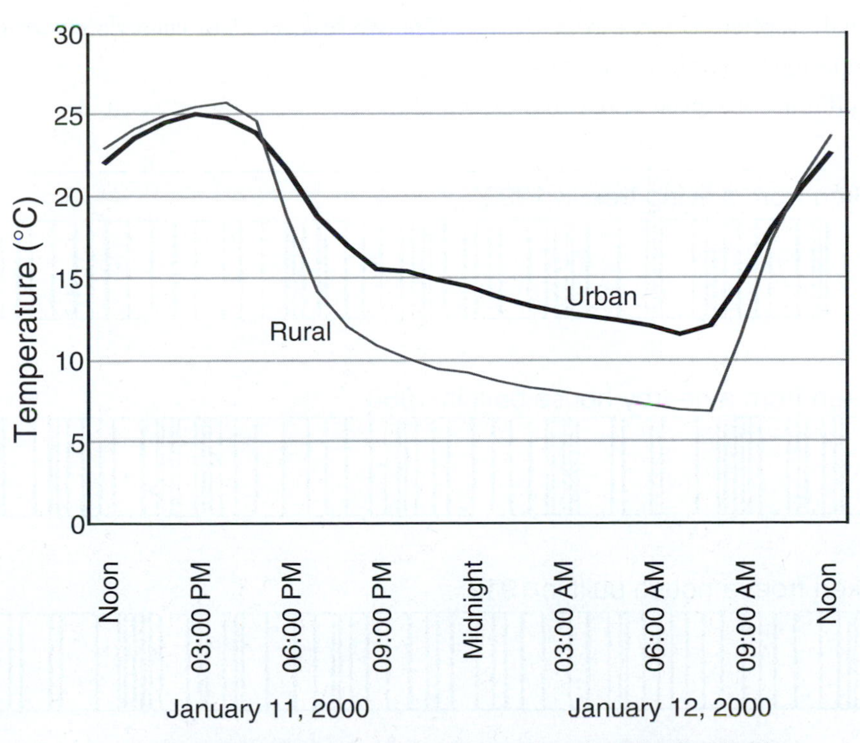

Figure 15-11. Rural vs. urban temperatures in Orlando, Florida.

Reconstructing Past Climate

Many of the physical mechanisms producing climate change act over long time periods and are, therefore, not evident in the relatively short instrumental climate record encompassing approximately the last 100 years. To uncover the details of past climate we must reconstruct the climate record using proxy data. Such reconstruction relies on the sensitivity of certain phenomena to climate. Plant and animal species, for example, respond to certain climate conditions, and their thresholds and their fossil record can indicate the presence of these conditions.

Tree-ring analysis has been used to reconstruct the climate over decades, centuries, and even millennia. It assumes that annual tree growth is limited or augmented by a particular climatic variable—such as summer temperature or winter and spring precipitation. For example, a narrow annual growth ring might indicate a dry year, a wide ring a wet year.

Tree ring analysis often requires cross-dating, i.e., comparing two or more cores to confirm the conditions of a given year. Cross-dating can also be used to extend the climate record into the past. Consider the three tree ring cores in Figure 15-12, all taken from an area where precipitation affects annual tree growth. The top core was taken in December 1990 from a living tree, and its outermost ring represents the growth in that year. Core 2 is from a beam of a house built in 1950. Core 3 is taken from a house built in 1912. The cores are read from right to left, so that the extreme left represents the oldest portion of each core.

15. *Examine the first core and label the newest ring 1990. Then mark every tenth ring, labeling them 1980, 1970, 1960, etc. Match the rings in the second core to the overlapping years in core 1 and continue marking every tenth ring, labeling the appropriate years. Do the same for the overlapping years between cores 2 and 3 to extend the record further into the past.*

16. *Which decades have unusually dry conditions?*

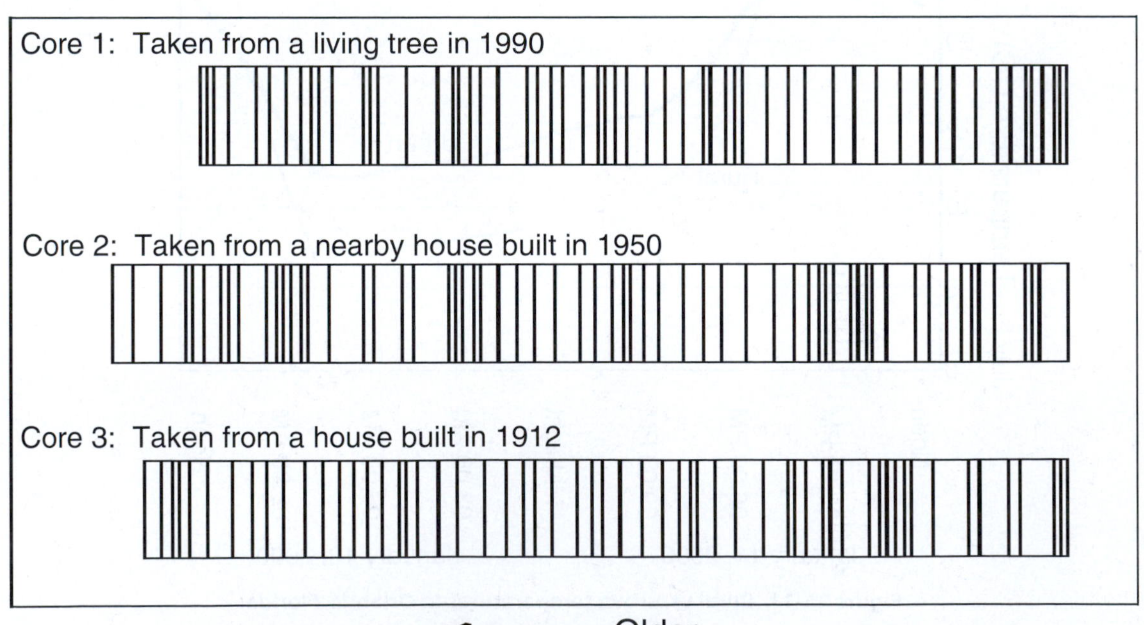

Figure 15-12. Tree ring cores.

Data from an Iowa tree ring study provide an example. Two researchers, Daniel Duvick and T. J. Blasing, cored white oak trees in south-central Iowa and correlated their annual ring width with precipitation during the preceding 12 months. Notice the relationship between precipitation and annual growth for the period 1880–1980 (Figure 15-13).

17. *Is the relationship stronger during wet years or dry years? (Hint: Circle the four driest seasons and the four wettest seasons and examine the corresponding ring width. Do the driest years correspond to the most narrow rings? Do the wettest years correspond to the widest rings?)*

Although tree rings provide detailed records for relatively recent climate history, many factors causing climatic change operate at longer time scales. The advance and retreat of glaciers, for example, have periods of tens of thousands of years or more. Fortunately, ocean fossils and ice cores preserve climate records at these longer time scales. Scientists have used two oxygen isotopes, ^{18}O and ^{16}O, found in ocean fossils as evidence of climatic change. Since ^{16}O evaporates more readily than ^{18}O, oceans are richer in ^{18}O during glacial advance, when water moves from oceans to continental glaciers. The shells of microorganisms produced during these periods preserve the ocean's higher ^{18}O concentrations. Ice cores from Greenland and Antarctica also preserve evidence of past climate or factors that influence climate. Air bubbles trapped in the ice, for example, can reveal past concentrations of greenhouse gases.

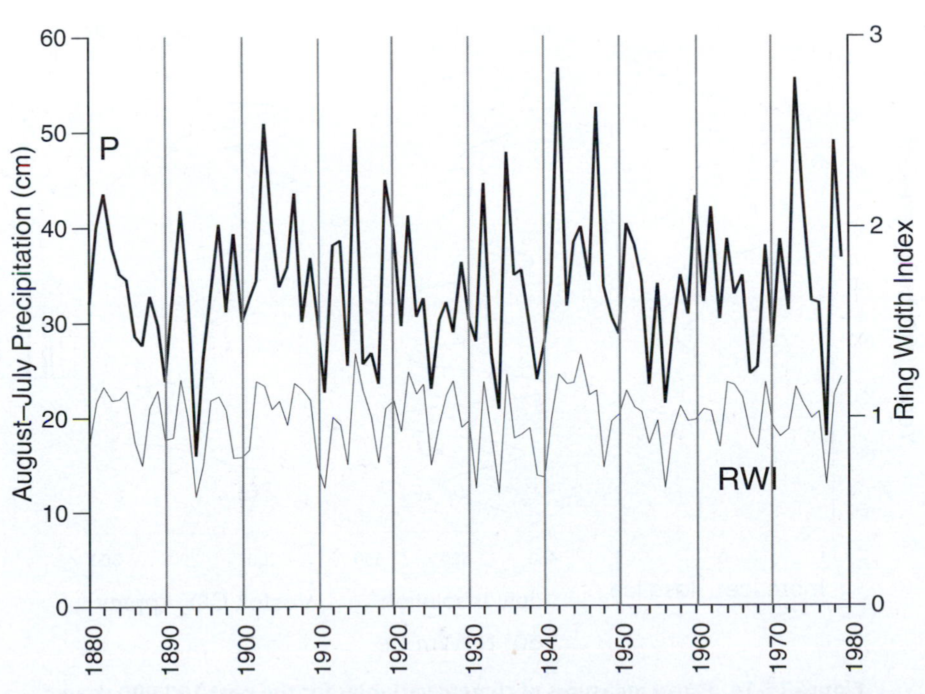

Figure 15-13. Tree-ring analysis: precipitation (dark solid line) vs. tree ring width (light dashed line).

Long-term variations in the path, tilt, and precession of the earth's orbit (Milankovitch cycles) strongly influence glacial advance and retreat. Glaciers retreat when these cycles combine to amplify the seasons in the Northern Hemisphere, creating relatively warm summers and relatively cool winters. Glaciers advance when the seasons are moderated—i.e., have relatively cool summers and relatively warm winters.

Figure 15-14 shows estimates of ice volume, CO_2 concentrations, and June insolation at 60° N latitude for the past 160,000 years. Use it to answer questions 18–22.

18. *When was the last time we had as little ice as we have today?*

19. *Examine the last glacial advance, 20,000 years ago. Was June insolation relatively high or relatively low at this time? How would a relatively cool summer contribute to glacial advance?*

20. *How could the CO_2 concentrations 20,000 years ago contribute to a global temperature that is cooler than today's?*

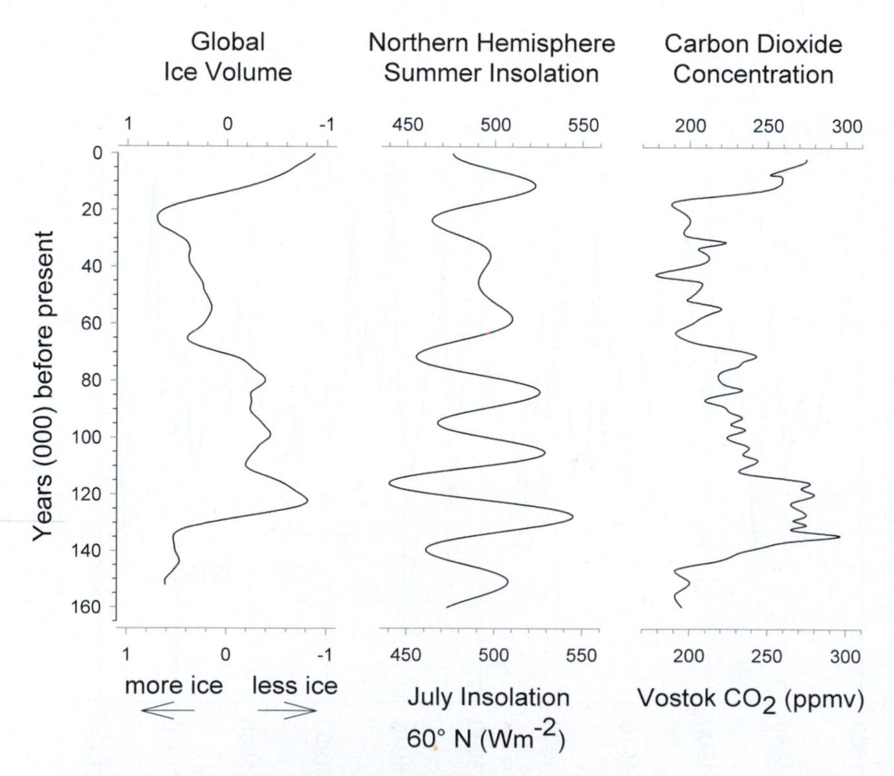

Figure 15-14. Proxy measures of climate variables for the past 160,000 years.

21. *Scientists examining paleoclimate records are often curious about which variables lead and which lag. How does the timing of June insolation minima relate to glacial ice volume at 135,000, 65,000, and 20,000 years before present? Which variable leads and which lags?*

22. *How is the relative timing of changes in ice volume and CO_2 more complicated?*

Optional Exercise: The Climate Record in Your Area

Examine the instrumental climate record for your area. You may want to collect data from the National Oceanic and Atmospheric Administration (NOAA) Climatological Data Series, which is found in many government documents libraries and is also available from various Internet sources (e.g., http://cdiac.esd.ornl.gov/r3d/ushcn/ushcn.html). Does the instrumental record show any special patterns over time? Can you make any general statements about climatic variability and change for your area?

You may be interested in the local climate during times preceding the instrumental record. Perhaps there are resources in a historical library such as records of missionaries, traders, or early settlers that could help you to reconstruct the climate. Often, diaries provide information about general temperature and precipitation patterns of the past.

Can you think of some other way that you could learn about the past climate of a particular area?

Review Questions

What challenges face climatologists interested in reconstructing the climate of the past?

What is the basis for using tree-ring analysis to reconstruct the climate? What are the potential advantages and disadvantages of this method?

Past climates could provide clues to future changes, but how does predicting the future climate become more complicated than simply extrapolating from past records?

Lab 16

Simulating Climatic Change

Introduction

Climate models allow us to simulate the response of the earth's climate to changes in particular environmental factors. In this lab, you will use the output from a variety of climate models to examine how these factors might change the global climate.

Climate Models and Radiative Forcing

While mathematical models cannot capture all the complexities of atmospheric processes, they do allow scientists to conduct "experiments" that estimate the climate's sensitivity to changes in such factors as solar output, volcanic activity, the earth's orbit, and greenhouse gases. These factors result in *radiative forcing*.

Radiative forcing is defined as the change in *net radiation* at the tropopause. (Recall that net radiation is the difference between incoming and outgoing radiation measured in W m^{-2}.) The change is usually calculated relative to "unperturbed" values, which are defined by the Intergovernmental Panel on Climate Change (IPCC) as those prevailing at the beginning of the Industrial Revolution (ca. 1750).

A positive forcing (an increase in net radiation) tends to warm the troposphere, while a negative forcing (a decrease in net radiation) tends to cool it. Note that a positive forcing may result from decreases in outgoing radiation as well as from increases in incoming radiation and that a negative forcing may result from increases in outgoing radiation as well as decreases in incoming radiation.

One type of climate model is an energy balance model. Energy balance models show how surface air temperature responds to short- and long-wave radiation fluxes. Output from a one-dimensional energy balance model developed by James E. Burt provides the basis for our examination of how solar variability, volcanic activity, and orbital changes affect those fluxes.

Solar Variability

Evidence suggests that solar output varies through time and that such variability could lead to climatic change. In the previous lab, you saw estimates for historical solar variability; now consider Figure 16-1, which shows simulated global surface temperature response to changes in the solar constant. The changes range from a 10% decrease to a 10% increase from current conditions.

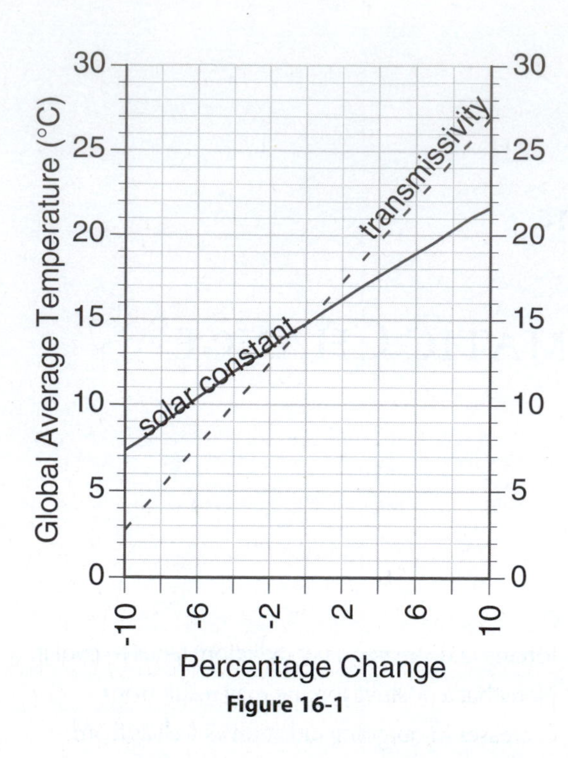

Figure 16-1

1. *What is the approximate average global surface air temperature simulated by the model without any alteration to solar constant?* ______________ °C

2. *What is the surface air temperature response to a 10% decrease in the solar constant?*

3. *Use Figure 16-1 to determine the simulated temperature response to a volcano that decreases atmospheric transmissivity by 1%.*

4. *How does this response compare with the temperature response to specific volcanoes observed in Lab 15 (Figure 15-6)?*

Optional Exercise

5. *Why would a 1% change in atmospheric transmissivity produce a greater surface temperature change than a 1% change in the solar constant? (Review the concept of the solar constant in the section on power in Appendix A.)*

Volcanic Activity

Volcanic activity also can affect the global climate. When a volcano explodes violently enough to spew sulphurous gases into the stratosphere, the resulting aerosols remain suspended, altering radiation fluxes. In particular, they scatter and reflect incoming solar radiation; that is, they decrease atmospheric transmissivity—the proportion of incident solar radiation that reaches the earth's surface when the sun is directly overhead.

Variations in the Earth's Orbit

In an attempt to explain certain climatic changes a Yugoslavian astronomer, Milutin Milankovitch, calculated variations in three components of the earth's orbit—eccentricity, precession, and obliquity. These variations affect seasonal and latitudinal receipt of solar radiation, and they partly explain periods of glacial advance and retreat.

6. *How would a decrease in the tilt of the earth change the intensity of the seasons in the mid-latitudes?*

7. *How would the intensity of the seasons change in the northern mid-latitudes if the perihelion changed from January 3 to some time near the Northern Hemisphere's summer solstice?*

Compare the earth's orbital eccentricity, precession, and tilt between current conditions and those 25,000 years ago (Table 16-1).

8. *Given the eccentricity and tilt 25,000 years ago, what differences in seasonality would you expect from the present?*

Table 16-1 also shows the simulated average global temperature 25,000 years ago and at present. The difference includes the influence of 35% lower carbon dioxide concentrations.

9. *What is the global average temperature difference between 25,000 years ago and present?*

The global temperatures shown in Table 16-1 are not uniform across the globe. Figure 16-2 shows simulated average monthly temperatures at 55° N latitude.

10. *During which season do you find the greatest difference between the present temperature and that 25,000 years ago at 55° N latitude? How might the conditions during this season have favored glacial ice there 25,000 years ago?*

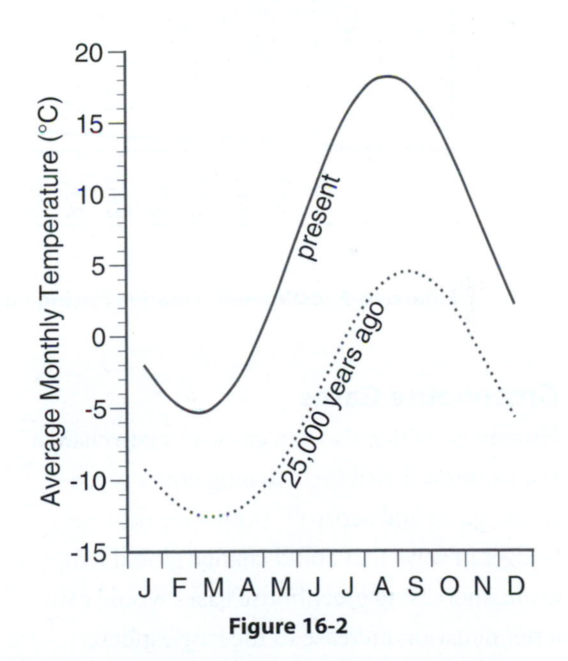

Figure 16-2

Table 16-1

	Present	25,000 yrs BP
Eccentricity	0.017	0.018 (+5.9%)
Tilt	23.44°	22.40° (–1.04°)
Perihelion	January 3	November 9
CO_2 Concentration	380 ppm	225 ppm
Global Temperature	15.2°C	8.8°C

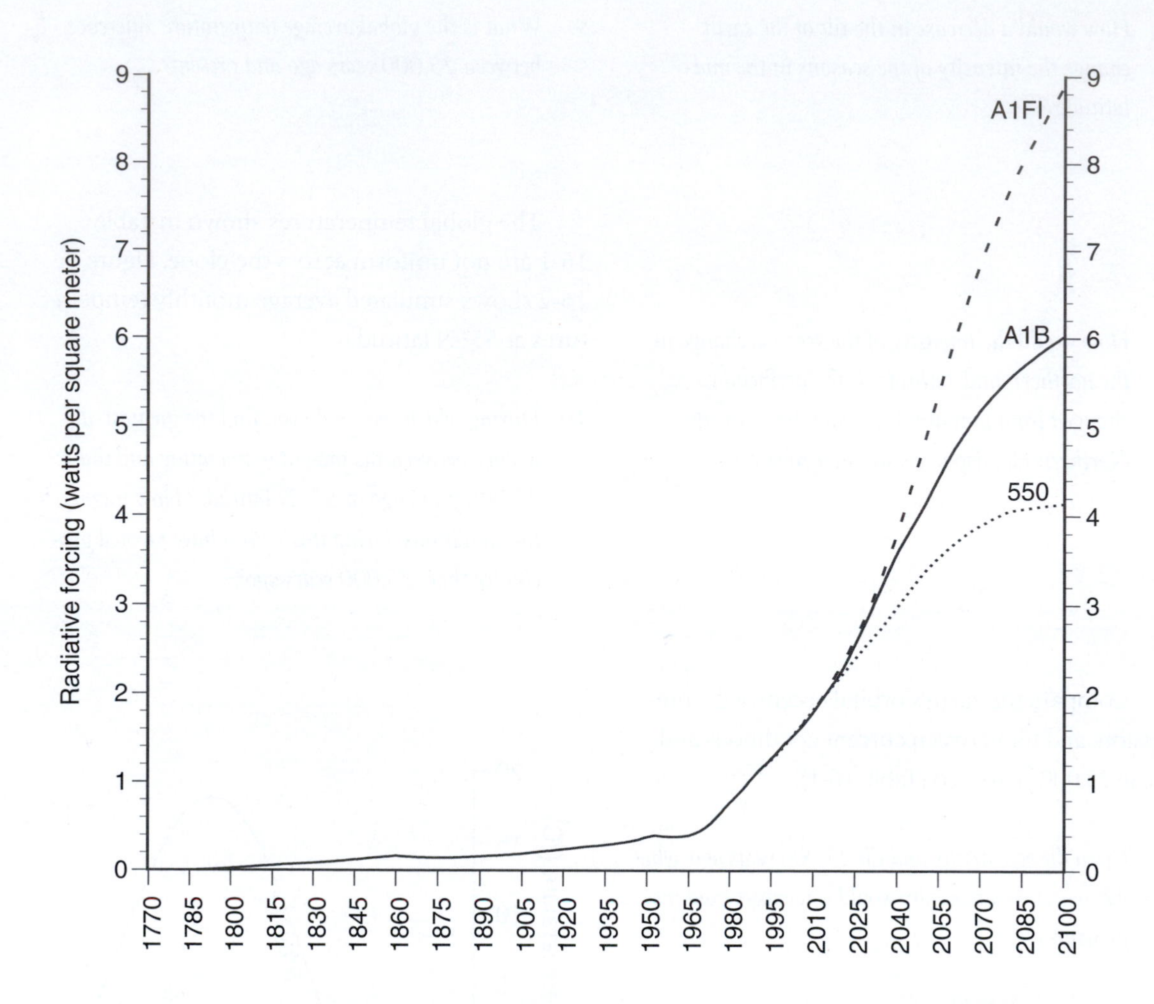

Figure 16-3. Estimated radiative forcing due to greenhouse gases (minus aerosol effect).

Greenhouse Gases

Human activities also can cause climate change. For example, fossil fuel burning emits greenhouse gases and aerosols. Both alter the energy budget in ways that could change global temperature. Increasing greenhouse gases would cause a net radiation increase to the troposphere, warming this layer. Many aerosols block incoming solar radiation, and their increase would cause net cooling. Since aerosols serve as condensation nuclei, they also influence cloud formation.

Figure 16-3 shows estimates of the radiative forcing due to anthropogenic activity. The data are derived from the Model for the Assessment of Greenhouse-Gas Induced Climate Change (MAGICC), which includes a gas-cycle model to estimate the positive and negative forcing from fossil fuel burning, changes in land-use, and aerosol emissions. This forcing drives an energy-balance model that predicts global temperature change. The graph shows estimates from 1770 to 1990, and projections from 1990 to 2100.

11. *Describe the magnitude of radiative forcing in the period before 1990 compared to projections from 1990 to 2100.*

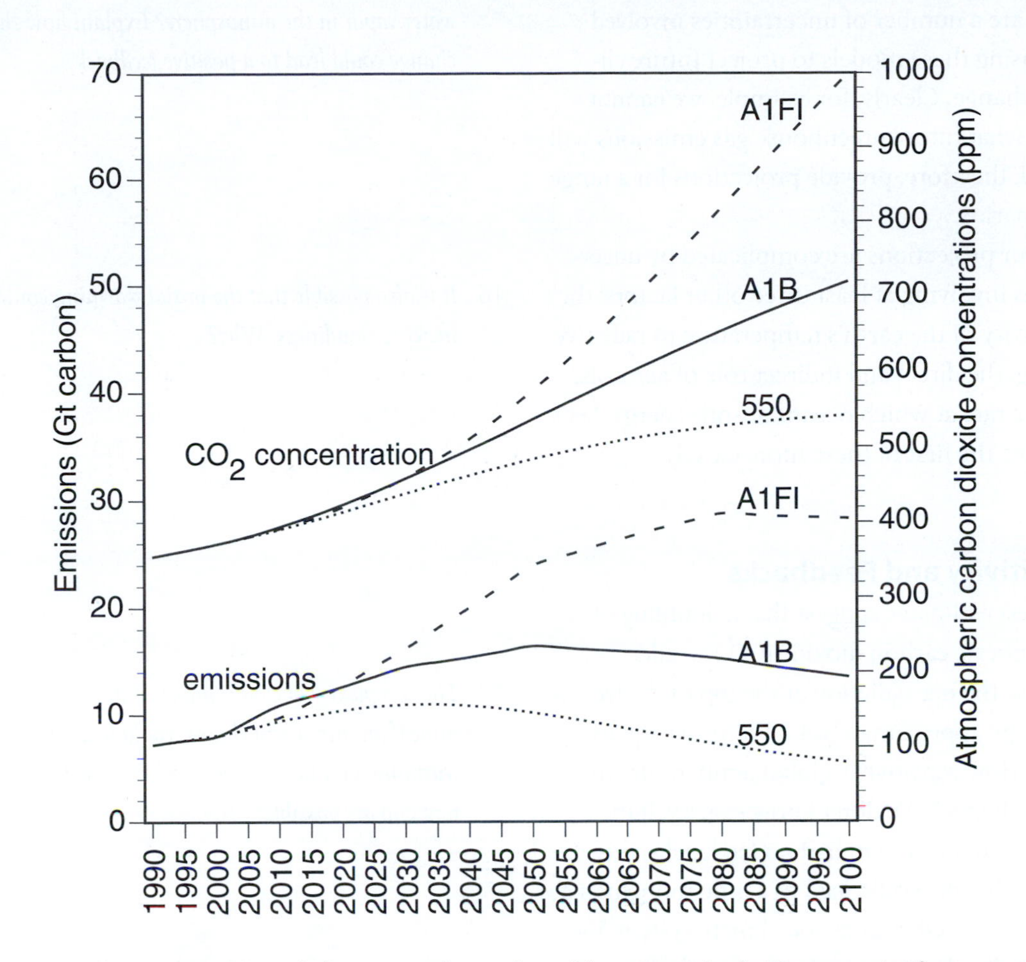

Figure 16-4. CO_2 emissions scenarios and resulting atmospheric CO_2 concentrations.

Figure 16-3 has three different lines in the 21st-century, each based on the estimated values of greenhouse gas concentrations shown in Figure 16-4. They represent very different scenarios. That labeled "A1FI" assumes intense future fossil fuel use; the "A1B" scenario assumes fossil fuel use gives way to alternative energy sources, and the "550 stabilization" scenario assumes immediate reduction strategies to stabilize the concentration of atmospheric carbon dioxide at 550 parts per million.

12. *To stabilize atmospheric carbon dioxide at 550 ppm, approximately when must emissions reductions begin? When do emissions peak in the other two scenarios?*

13. *Describe the relationship between emissions and concentrations of carbon dioxide for each scenario.*

Uncertainties about Climate Change

There are a number of uncertainties involved with using these models to project future climate change. Clearly, for example, we cannot know what future greenhouse gas emissions will be and, therefore, provide projections for a range of scenarios.

Our projections are complicated by uncertainties involving at least three other factors: the sensitivity of the earth's temperature to radiative forcing, the direct and indirect role of aerosols, and the rate at which oceans absorb energy. Let's examine the first of these more closely.

Sensitivity and Feedbacks

Our best estimates suggest that a doubling of atmospheric carbon dioxide (CO_2) would increase the net radiation at the top of the troposphere by approximately 4.2 Watts per square meter. How sensitive is global temperature to such a change? We don't know exactly. If the earth's climate system had no feedbacks, this energy change would lead to a little more than a 1°C increase. Of course, the climate system does have feedbacks. An initial warming due to an additional 4.2 Watts per square meter could lead to feedbacks associated with snow and ice, water vapor, and cloud cover. These feedbacks could either amplify the initial change (positive feedbacks) or regulate the effects of the initial change (negative feedbacks).

14. *How might an initial warming alter snow and ice on the planet? Would this change amplify or regulate the initial change? Explain your answer.*

15. *How could an initial warming alter the amount of water vapor in the atmosphere? Explain how this change could lead to a positive feedback.*

16. *It is also possible that the initial warming could increase cloudiness. Why?*

17. *The complexity of clouds makes it difficult to determine if an increase in clouds would lead to a net warming or a net cooling. Explain how both phenomena are possible.*

In its 2007 Fourth Assessment Report, the Intergovernmental Panel on Climate Change examined scientific evidence for climate feedbacks from observations and climate models. The IPCC concluded that when best estimates of feedbacks are considered, the global average temperature increase resulting from an instantaneous doubling of carbon dioxide (sometimes referred to as the equilibrium climate sensitivity) falls in the range of 2°–4.5°C, with a value of 3°C most likely.

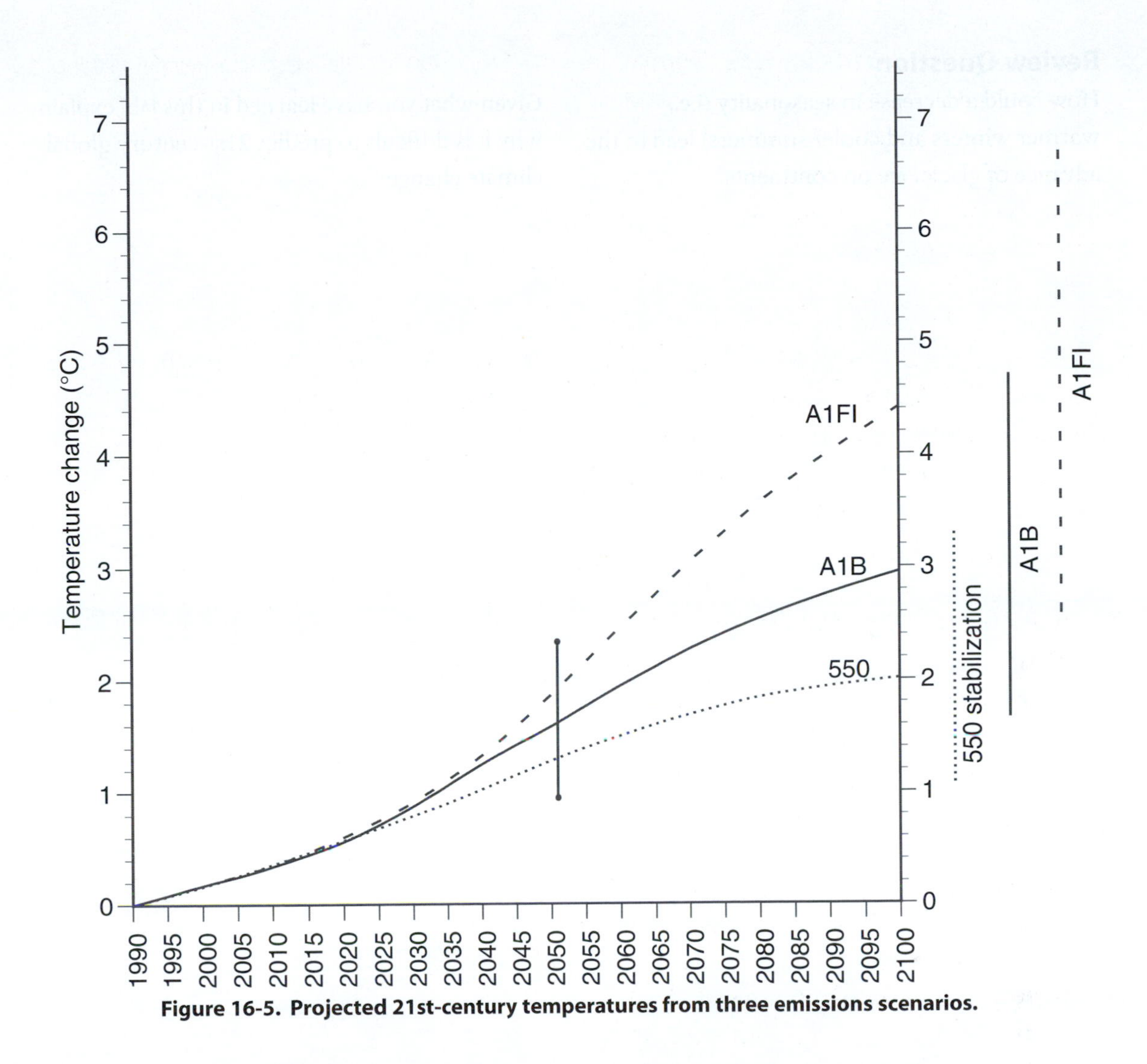

Figure 16-5. Projected 21st-century temperatures from three emissions scenarios.

Figure 16-5 shows global temperature change projections based on the three emissions scenarios mentioned previously. Each line represents estimated change assuming mid-level estimates of climate sensitivity, aerosols concentration, and ocean uptake of energy. The bars to the right of the diagram show the range of global temperature change by 2100 that would result from low (1.5°C) to high (4.5°C) climate sensitivity values. Likewise, the solid bar at year 2050 shows the range of projected temperature change for the A1B emissions scenario.

18. *According to these simulations, how does the magnitude of temperature response to greenhouse gas increases compare to some of the other radiative forcings described in this lab?*

19. *To what degree would emissions reductions influence global temperature in the next 25 years?*

20. *Describe the importance of knowing the emissions path and the earth's climate sensitivity to projected global temperature change by the end of the 21st century.*

Review Question

How could a decrease in seasonality (i.e., warmer winters and cooler summers) lead to the advance of glacial ice on continents?

Given what you have learned in this lab, explain why it is difficult to predict 21st-century global climate change.

Appendix A

Dimensions and Units

Materials Needed

- calculator

Systems of Measure

To understand the features and processes of the atmosphere, we must consider how meteorological variables such as energy, temperature, and pressure are measured. Meteorological variables can be expressed in terms of the following fundamental dimensions:

a. Time (*T*)
b. Length (*L*)
c. Mass (*M*)
d. Temperature (*D*)

Several systems have been developed to express dimensions in units. The most common systems are the cgs (centimeter-gram-second), SI (Système International), and English systems. The fundamental dimensions in each system are expressed in the following units:

	cgs	SI	English
Time	second	second	second
Length	centimeter	meter	foot
Mass	gram	kilogram	pound
Temperature	kelvin	kelvin	°Fahrenheit

Dealing with Units

We frequently need to work with units of different dimensions (e.g., meters, calories, seconds) in meteorological calculations. This is done by treating such units as variables. Compare, for example,

$$\frac{75x}{3y} = \frac{75}{3} \bullet \frac{x}{y} = 25\frac{x}{y}$$

with

$$\frac{75\ meters}{3\ seconds} = \frac{75}{3} \bullet \frac{meters}{seconds} = 25\ m\ s^{-1}$$

In the second equation, meters and seconds are treated as variables. The symbols 75 meters and 3 seconds represent the products of 75 times meters and 3 times seconds, respectively. The symbol meters/seconds represents a quotient of meters divided by seconds. We may express units appearing in the denominator with negative exponents. Thus, in the above example, our solution may be written 25 meters · second^{-1} or abbreviated as 25 m s^{-1}.

Metric units are most commonly used in meteorology; however, because English units are used in the United States, it is sometimes necessary to convert between the two systems. Although many sources provide conversion tables, and calculators often perform conversions, there are instances when conversion between specific units may not be readily available. There are standard methods for such cases.

When converting units from one system to another we use the properties that anything multiplied by 1 equals itself and, conversely, anything divided by itself is equal to 1. Thus, 12 inches divided by 12 inches equals 1. Similarly, 12 inches divided by 1 foot also equals 1, as does 1 inch divided by 25.4 millimeters. There are occasions when multiplication by unity may be used several times in converting values. For example, we may want to express 12.5 feet in meters. Using Table A-1 as a reference, we might make this conversion in the following fashion:

A. Find a conversion between metric and English units of length from Table A-1 (1 inch = 25.4 mm).

B. Multiply by a series of relevant unity values to calculate the length in the desired units (meters) and "cancel" unwanted units:

$$12.5\ \mathit{feet} \bullet \frac{25.4\ mm}{1\ inch}$$

(English/metric conversion)

$$12.5\ \mathit{feet} \bullet \frac{25.4\ mm}{1\ \not{inch}} \bullet \frac{12\ \not{inches}}{1\ \not{foot}}$$

(cancel feet, inches)

$$12.5\ \not{feet} \bullet \frac{25.4\ \not{mm}}{1\ \not{inch}} \bullet \frac{12\ \not{inches}}{1\ \not{foot}} \bullet \frac{1\ meter}{1000\ \not{mm}}$$

(cancel mm)

We are left with:

$$12.5 \bullet \frac{25.4}{1} \bullet \frac{12}{1} \bullet \frac{1\ m}{1000} = \frac{3810\ m}{1000} = 3.81\ m$$

Notice that each factor we multiply by 12.5 (e.g., 25.4 mm/1 inch, 12 inches/1 foot, and 1 meter/1000 mm) equals 1. In addition, all units except meters cancel, since dividing feet by feet, mm by mm, or inches by inches results in a quotient of 1.

Variables Important to Meteorology

Length

Dimension: Length (*L*)

1. *Convert the typical altitudes of:*

   a. geostationary satellites (36,000 kilometers)

   ____________ *miles*

   b. cruising jet airplanes (37,000 feet)

   ____________ *meters*

   c. Most of earth's weather results from processes occurring below this cruising height. What is the ratio of this height to the earth's radius (6400 km)?

2. *In the mercury barometer (developed by Evangelista Torricelli), air pressure forces mercury up a glass tube. At sea-level pressure, the mercury is pushed to 760 mm. This height is equal to ____________ inches of mercury.*

Table A-1

Length

1 inch	=	25.4 millimeters
1 statute mile	=	1.6 kilometers
1 nautical mile	=	1.1516 statute miles
1° latitude	=	111 kilometers

Area

1 square yard (yd^2)	=	0.8361 square meter (m^2)
1 square inch ($in.^2$)	=	6.45 square centimeters (cm^2)
1 square meter (m^2)	=	1.196 square yards (yd^2)
1 square meter (m^2)	=	10,000 square centimeters (cm^2)
1 square centimeter (cm^2)	=	0.155 square inch ($in.^2$)

Volume

1 milliliter (ml)	=	1 cm^3

Force

1 newton	=	0.2248 pound (lb)

Pressure

1 Newton per square meter	=	1 pascal (Pa)
1 mm mercury	=	133.32 pascals
1 pound per square inch	=	6895 pascals

1 bar (1000 mb) = 100,000 pascals = 100 kilopascals (kPa) = 1000 hectopascals (hPa)

Standard sea-level pressure, or 1 atmosphere, can be expressed by any of the following units:

1 atmosphere = 101325 pascals = 1.01325 bars = 1013.25 millibars

1 mm mercury = 1.333 mb

Energy

1 joule (J)	=	0.2389 calories
1 British thermal unit (BTU)	=	1055 joules = 252 calories
1 kilowatt-hour (kWh)	=	3,600,000 joules

Temperature

° Celsius	=	5/9 • (°Fahrenheit – 32.0)
° Celsius	=	kelvins – 273.15

Area

Dimensions: Length • Length (L^2)

3. *Convert the following:*

10 in.^{2} = ____________ cm^2

20 m^2 = ____________ ft^2

(Hint: If you imagine 20 m^2 as a perfect square, you can determine the length of one side of that square by simply calculating the square root of the area.)

Velocity

Dimensions: Length • Time^{-1} (LT^{-1})

6. *If a strong west wind blows at 35 nautical miles per hour (35 knots), how fast is the wind moving:*

in miles per hour? ____________ mph

in kilometers per hour? ____________ kph

in meters per second? ____________ m s^{-1}

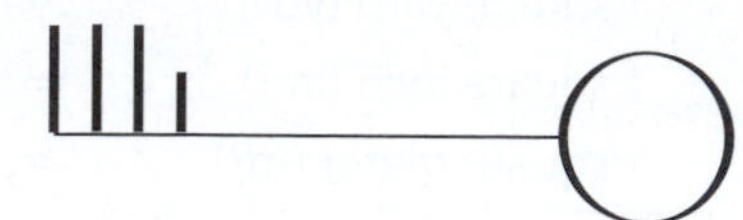

Figure A-2. Westerly wind, 35 knots.

Volume

Dimensions: Length • Length • Length (L^3)

4. *What is the volume of a box with the following dimensions:*

width: 2 m

height: 2 m

depth: 2 m

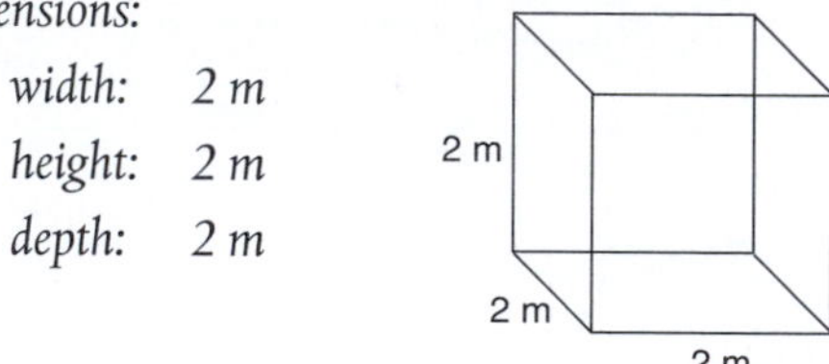

Figure A-1

____________ m^3 = ____________ ft^3

Acceleration

Dimensions: Length • Time^{-2} (LT^{-2})

Acceleration is defined as a change in speed or direction with time. In SI units, acceleration is expressed as meters per second squared (m s^{-2}). There are several types of acceleration important in meteorology, including gravitational and Coriolis acceleration. Acceleration due to gravity equals 9.8 m s^{-2}. That is, every second the velocity of a freely falling object increases by 9.8 m s^{-1}.

Density

Dimensions: Mass • Length^{-3} (ML^{-3})

Density is equal to mass divided by volume. In SI units, it is most often expressed as kg m^{-3}. Air density at sea level generally will vary between 1.1 and 1.3 kg m^{-3}, depending on air temperature and pressure.

5. *If the box in Figure A-1 had a sea-level air density of 1.23 kg m^{-3}, how much would the air within the box weigh? ____________ kg*

Force

Dimensions: Mass • Length • Time^{-2} (MLT^{-2})

Atmospheric forces cause vertical and horizontal movement of air. In SI units, force is expressed in newtons (kg m s^{-2})—the force necessary to accelerate a 1-kilogram mass 1 meter per second every second. Newton's second law expresses force as:

Force = mass • acceleration

7. *What is the gravitational force on the air in the box described in Figure A-1?*

________ kg m s^{-2} = ________ newtons (N)

Pressure

Dimensions: Length^{-1} • Mass • Time^{-2} ($L^{-1}MT^{-2}$)

Pressure is an important variable in meteorology and is best understood as force exerted over a unit area. Pressure is expressed in SI units as newtons per meter squared (N m^{-2}), or pascals (1 N m^{-2} = 1 pascal).

8. *If the box in Figure A-1 rests on the earth's surface, it occupies an area of 4 m^2. What is the pressure exerted as gravity pulls on the air in the box?*
   ____________ *N* m^{-2} = ____________ *Pa*

Another unit of pressure is the bar. It was chosen to approximate the pressure of the atmosphere at sea level. Standard atmospheric pressure equals 101,325 pascals (Pa), and 1 bar = 100,000 Pa = 100 kPa = 1000 hPa. Pressure is commonly expressed on weather maps in millibars, i.e., one-thousandth of a bar (1 millibar = 0.001 bar).

9. *Convert the following:*
   1 millibar (mb) = ____________ *pascals (Pa)*

10. *Express standard sea-level pressure in millibars*
    ____________ *mb*
    and in kilopascals ____________ *kPa.*

11. *Convert the following: 102.4 kPa =*
    ____________ *mb* =
    ____________ *mm mercury* =
    ____________ *inches.*

Energy

Dimensions: Length^2 • Mass • Time^{-2} ($L^2 M T^{-2}$)

Energy measures the ability to do work and thus can be considered as the measure of a force acting over some distance. In the SI system, energy is expressed in joules. One joule (J) is equal to 1 newton acting over a distance of 1 meter (1 joule = 1 N m). Another commonly used unit of energy is the calorie, which is derived from the cgs system. It is common to examine energy over some area (e.g., joules per square meter or calories per square centimeter). In the cgs system, 1 calorie per square centimeter is equal to 1 langley. Because of the variety of units used to measure energy, it is often necessary to convert between systems. Consider, for example, conversion between langleys and joules per square meter. Substituting 1 calorie per square centimeter for 1 langley we have:

$$\frac{1\ \cancel{cal}}{\cancel{cm}^2} \cdot \frac{1\ J}{0.2389\ \cancel{cal}} \cdot \frac{10{,}000\ \cancel{cm}^2}{1\ m^2} = 41{,}858.5\ J\ m^{-2}$$

Power

Dimensions: Length^2 • Mass • Time^{-3} ($L^2 M T^{-3}$)

Power is an expression of energy per unit time. In the SI system, 1 joule per second = 1 watt.

Like energy, power is commonly expressed per unit area. The solar constant is an example of this measure: it is the rate of solar energy receipt on a hypothetical two-dimensional disk, perpendicular to the sun's rays, and ignoring the effects of the atmosphere (Figure A-3). The solar constant is approximately 1.96 cal cm^{-2} (langleys) per minute.

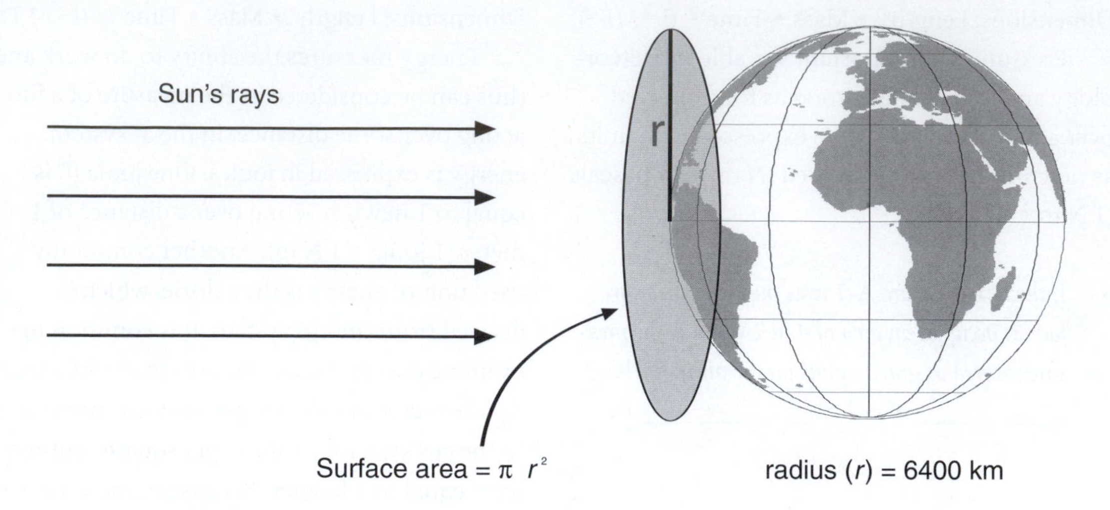

Figure A-3

12. Express the solar constant in watts per square meter.

____

Temperature

13. Select the maximum and minimum temperatures from a recent day and convert to °C and kelvins.

Maximum Temperature	*Minimum Temperature*
____________°F	____________°F
____________°C	____________°C
____________K	____________K

Appendix B

EARTH MEASURES

Materials Needed

- world atlas

Latitude and Longitude

Because of earth's near-spherical shape, it is convenient to measure locations on the earth in degrees. The circumference of the earth is equal to 360°. Latitude measures position relative to the widest bulge of the earth, the equator (0° latitude), and is expressed in degrees north or south (° N, ° S).

Since there are 90° in a quarter of a sphere (360° ÷ 4), the North and South Poles are located at 90° N and 90° S latitude, respectively.

1. *Find the latitude of the following locations:*

____________ *Equator*
____________ *Kansas City, MO, USA*
____________ *Arctic Circle*
____________ *Montreal, Canada*
____________ *Antarctic Circle*
____________ *Tropic of Cancer*
____________ *Mexico City, Mexico*
____________ *Tropic of Capricorn*
____________ *Bombay, India*
____________ *Cape Town, South Africa*

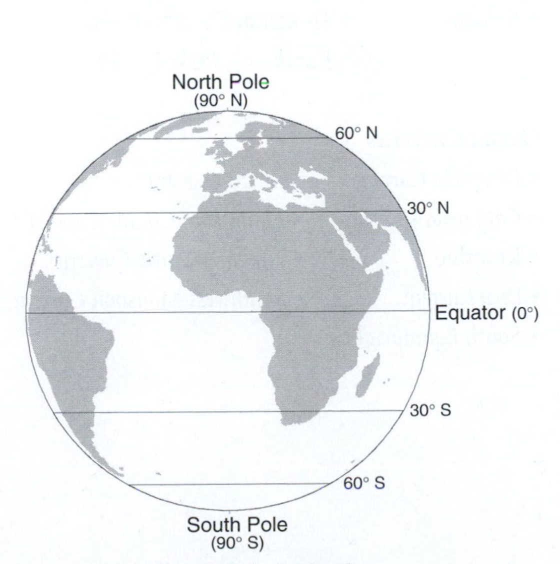

Figure B-1. Latitude.

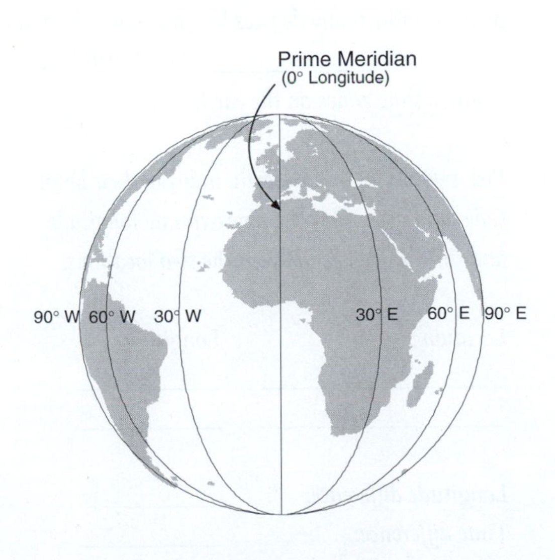

Figure B-2. Longitude.

Longitude measures position east or west of the *prime meridian*—a line passing from the North to South Pole, through Greenwich, England. Lines of longitude to the east of the prime meridian are expressed in degrees east (° E), and those to the west of the prime meridian in degrees west (° W). Halfway around the earth from Greenwich the lines of east and west longitude meet at 180°.

Both longitude and latitude can be broken down into 60th fractions, or minutes. Minutes are broken down into 60 seconds.

1° = 60 minutes (60')
1' = 60 seconds (60")

2. *Find the latitude and longitude of:*

____________ *Los Angeles, CA, USA*
____________ *Lima, Peru*
____________ *Baghdad, Iraq*
____________ *Moscow, Russia*
____________ *Melbourne, Australia*
____________ *Rome, Italy*
____________ *Nairobi, Kenya*
____________ *Atlanta, GA, USA*

3. *The earth makes a complete rotation on its axis every 24 hours. Since the earth's circumference is 360°, by how many degrees longitude does it rotate every hour?* ____________ *How does this value relate to time zones on the earth?*

4. *Pick two places on the earth, indicate their longitude, and determine the approximate longitude and time difference between the two locations.*

Location	*Longitude*
______________	________
______________	________

Longitude difference: ________
Time difference: ________

General Geography

Atmospheric processes are interconnected. What happens in one part of the world can affect the weather and climate in another part. Since we will make occasional reference to places around the world, it is essential that you review a bit of world geography. Study your atlas and locate the world's continents and oceans.

5. *Mark the following places on the world map provided.*

Continents

- *Africa*
- *Antarctica*
- *Asia*
- *Australia*
- *Europe*
- *North America*
- *South America*

Oceans

- *Arctic Ocean*
- *Atlantic Ocean*
- *Indian Ocean*
- *Pacific Ocean*

Mountain Ranges

- *Alps*
- *Andes*
- *Appalachians*
- *Cascades*
- *Caucasus*
- *Himalayas*
- *Rockies*
- *Urals*

Deserts

- *Arabian*
- *Atacama*
- *Gobi*
- *Great Victorian*
- *Kalahari*
- *Mojave*
- *Sahara*
- *Turkestan Deserts (Peski Karakumy, Peski Kyzyl Kum)*

Ocean Currents

- *Benguela Current*
- *Brazil Current*
- *California Current*
- *Equatorial Countercurrent*
- *Kuroshio*
- *North Atlantic Current*
- *Peru Current*
- *Southwest Monsoon Current*
- *South Equatorial Current*

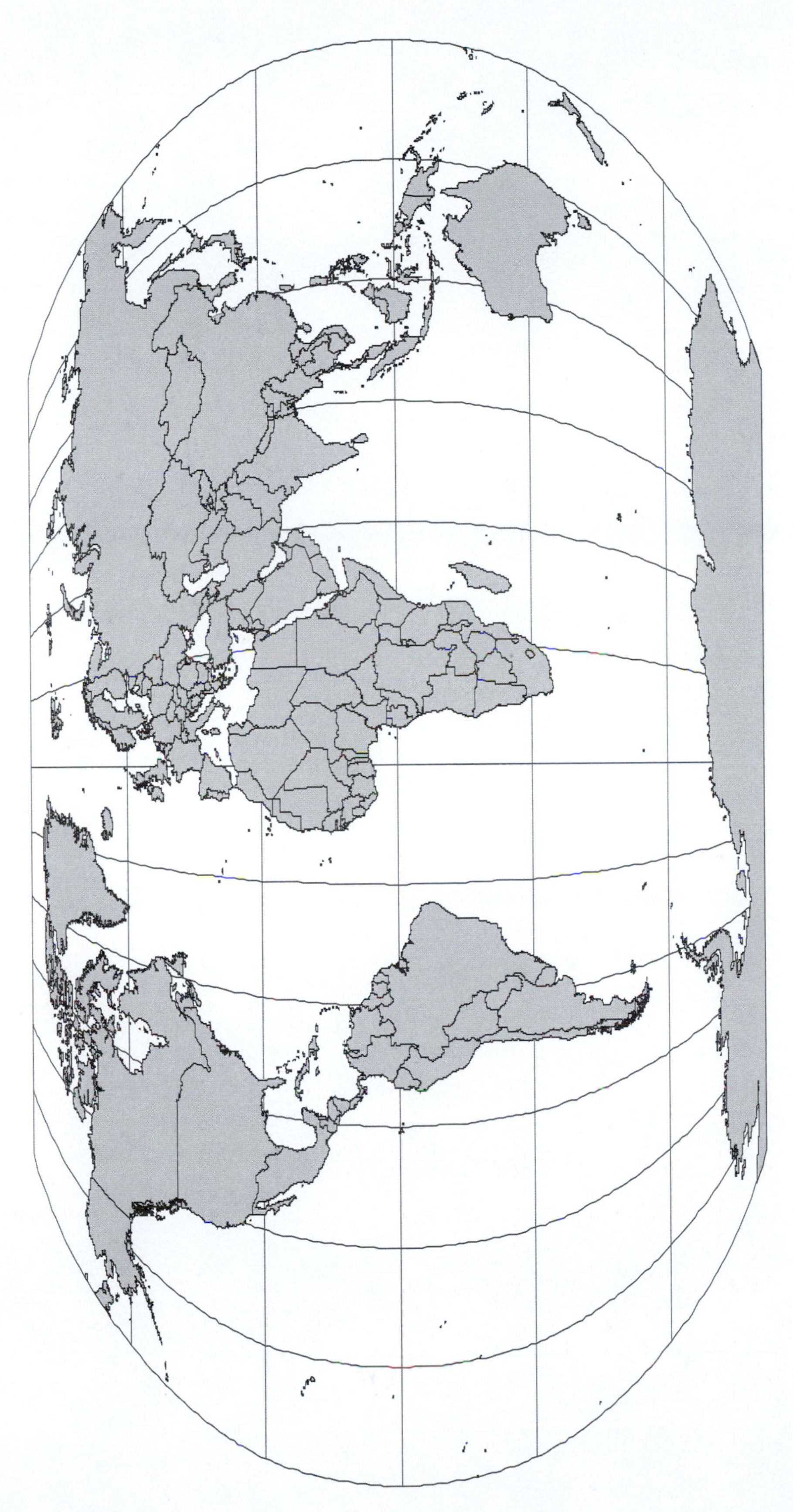

Appendix C

GeoClock

Materials Needed

- GeoClock, shareware that is available at http://home.att.net/~geoclock/.

Introduction

Why might a region be referred to as the "land of the midnight sun?" Why do places on the equator receive 12 hours of daylight every day of the year? This lab uses GeoClock, a computer program written by Joseph R. Ahlgren, to investigate how earth–sun geometry influences daylight hours.

Running GeoClock

When you begin, you will see a graphic similar to the one below. Take a minute or two to familiarize yourself with the map and the text at the bottom. The illuminated and dark portions of the earth correspond to the internal clock and calendar on your computer. For a given location, at the specified day and time, the text provides the sun's azimuth and elevation (solar elevation angle), and the latitude and longitude of the sun's direct rays. It also shows the time of sunrise and sunset for a specific location. Note that time is elapsing at a normal rate.

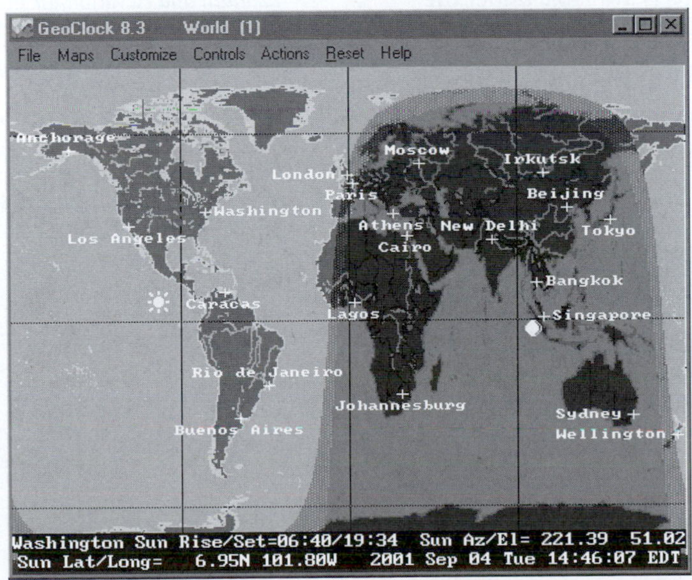

Figure C-1. GeoClock screen.

GeoClock Commands

GeoClock commands can be accessed using menus at the top of the screen or by using the keyboard. You will find most of the commands required for this lab under the "Maps" and "Controls" menus. Here are some keyboard shortcuts:

Time Control

T Allows you to set the date, time, and speed at which time elapses.

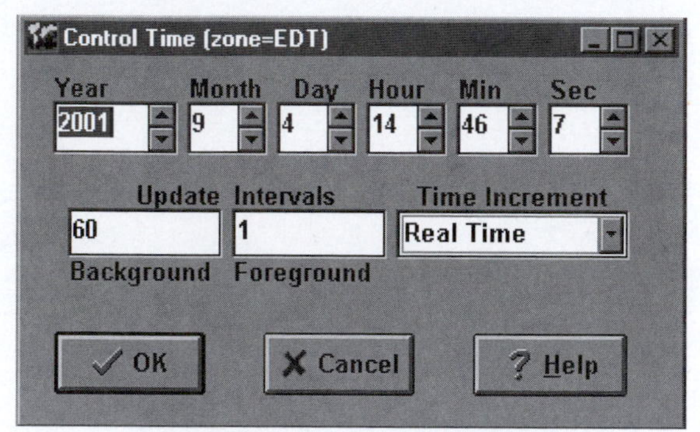

Figure C-2. Time control window. Accessed by simply typing T.

Map Selection

M Select a map from the list.

K Return to previous map display.

Special Commands

C Access a command window. A full list of commands is available under the "Help" menu.

V Access the VCR controls window, to run, stop, and reverse the sun's motion. Clicking the **T** button will also bring up the Control Time window.

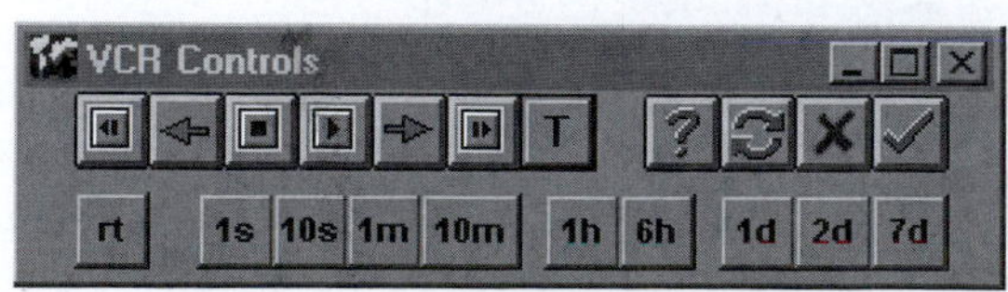

Figure C-3. VCR Controls window.

If you haven't already done so, start GeoClock.

Before you answer questions 1–5, you may want to change the "Primary City" shown on the bottom of the GeoClock screen to a place close to home. To do this, select **Customize MAP1** under the "Customize" pull-down menu. Then select **Change Primary**. You can then enter a city name with its latitude and longitude, or simply enter the ZIP code. You could also click on the **City List** button, scroll to a city of your choice, click on its name, and then click the **OK** button.

1. *What is the city shown at the bottom of the screen? What are the date and the time?*

2. *What are the latitude (solar declination) and longitude of the direct solar rays?*

3. *The "Sun Az/El " refers to the position of the sun. Azimuth tells you its direction (90° is east, 180° is south, and 270° is west). Elevation is the solar angle above the horizon. What is the current solar azimuth and elevation for the city shown?*

4. *When does the sun rise and set at the city listed? How many hours of daylight does it have on this date?*

5. *Solar noon is the midpoint in time between sunrise and sunset. At what time does it occur at this city on this date?*

6. *Besides the time, which variables at the bottom of the screen are changing? Explain why they are changing.*[*]

7. *How would you characterize the shape of the boundary between light and dark portions of the earth? Is a greater proportion of one hemisphere illuminated than of the other?*

To examine daylight hours at different latitudes, it is helpful to draw reference latitude and longitude lines on the globe. The commands below will allow you to draw lines of longitude at 15° intervals that divide the globe (360°) into 24 equal parts. Although these lines of longitude do not divide the world into actual time zones, they can be used to estimate daylight hours for various parts of the world, since each 15° of longitude represents 1 hour.

[*] If the time at the bottom of the screen is not changing, simply click the reset button at the top of the screen.

Relevant Commands:

- Access the command window by typing **C** (or choose: Controls | Command entry).
- Enter **glinc 15** in the command window and click the **OK** button.
- Enter **tlinc 15** in the command window and click the **OK** button.
- Click the **Close** button to exit the command window.

8. *London is located at 0° longitude, Los Angeles at 118° 15' W longitude. What is the difference in longitude between these two cities?* __________°
   Divide this number by 15° to estimate the time difference between the two cities: __________*hours.*

   What is the difference in longitude between Washington, D.C. (72° 2' W), and Beijing, China (116° 25' E)? __________°
   What is the approximate time difference between the two cities? __________*hours.*

9. *Use the graphic to estimate the daylight hours at three different latitudes on four different dates*[†]*:*

Daylight Hours

Latitude	________	________	________
Today			
June 21			
Dec. 22			
Mar. 21			

[†] ***Relevant Commands*** *(to change the date)*:

- Type **V** to open the VCR Controls window (or choose: Controls | VCR Controls).
- Click **T** in the VCR Controls window (you may want to drag the "Control Time" window away from the main viewing area).
- Enter the appropriate date.
- Click the **OK** button.

We have examined daylight hours for several dates and places but have not displayed gradual seasonal changes. In this section, you will view changes in daylight hours at one-day intervals. Read through the questions and Relevant Commands before beginning this section.

Relevant Commands:
In the VCR Controls box:

- Click **T** in the VCR Controls window and change the "update intervals" in the foreground from 1 to **0.1**.
- Click the **fast forward** button.
- Choose **1d**. (Note that each time the screen is updated, the date changes by one day. Since you are examining the same time each day, the sun remains at the same longitude but changes latitude each day.)
- Toggle between the **pause** and **fast forward** buttons as needed.

10. *Observe the pattern of illumination and the date at the bottom of the screen as the seasonal cycle progresses. When do the number of daylight hours increase in the Northern Hemisphere? When do they increase in the Southern Hemisphere?*

11. *Watch the changes in the pattern of illumination and darkness through the course of a year. Does this pattern, and hence daylight hours, change most rapidly near the solstices or near the equinoxes?*

Diurnal Changes

This section shows changes in the illumination of the earth during the course of a day as the earth rotates on its axis. You will first change the time of year and then adjust the time rate to half-hour intervals. Observe the patterns of daylight change during the course of the day and, more gradually, during the course of the year.

Relevant Commands:
In the GeoClock window:

- Choose the **Space—N America** map (type **M** or choose: Maps | Map List).
- Open the VCR Controls box.
  –Click **T** in the VCR Controls window and set the foreground update intervals to **1**.
  – Click the **reset** button.
  – Click the **fast forward** button.
  – Click the **1h** button.

A View from the Poles

In this section you will examine daylight hours during each of the solstices and from the perspective of both poles. Read through the questions and the Relevant Commands before continuing.

Relevant Commands:
In the GeoClock window:

- Enter **M** to select the South Pole from the map list.
- Enter **M** to select the North Pole map.
- Enter **K** to switch between the last two maps.
- Enter **T** to change the dates accordingly.

12. *What can you say about daylight hours in the South Polar regions around the December solstice? What can you say about daylight hours in the North Polar regions around the December solstice?*

13. *How are daylight hours in polar regions different around the June solstice?*

Examine seasonal changes at the Poles using the following commands:

Relevant Commands:

In the VCR Controls box:

- Click the **fast forward** button and change the animation time to **1d**.

In the GeoClock window:

- Press **K** to toggle between the North Pole and South Pole maps.
- Pause and start the animation as needed.

14. *Toggle between the North and South Pole maps and observe how daylight hours change seasonally at the poles. Freeze the screen when the circle of illumination cuts through the North or the South Pole. When does this occur?*

15. *Freeze the screen when the Antarctic region has its longest period of daylight. On which day does this occur?*

The Direction of Sunrise and Sunset

As a final exercise you will examine the patterns of sunrise, sunset, and daylight hours across the United States on the two solstices and on the March equinox. Again, you should read through the questions and Relevant Commands first.

Relevant Commands:

In the GeoClock window:

- Enter **M** to change the map to United States (48).

In the VCR Controls window:

- Enter **T** to adjust the dates; change foreground update intervals to **0.05**.
- Click the **fast forward** button and select a time increment of **10m**.
- Use the VCR controls as needed.

16. *For a given line of longitude, does the sun rise and set first in the northern or southern United States? (Circle the correct answer.)*

Rises First

June 21	*Northern USA*	*Southern USA*
Dec. 22	*Northern USA*	*Southern USA*
Mar. 21	*Northern USA*	*Southern USA*

Sets First

June 21	*Northern USA*	*Southern USA*
Dec. 22	*Northern USA*	*Southern USA*
Mar. 21	*Northern USA*	*Southern USA*

17. *Record the azimuth at sunrise and sunset for your primary city on the March equinox and on the solstices. On the figure below, place a mark indicating each sunrise and sunset direction for all dates. This is already done for the sunrise on the March equinox (when the sun rises directly from the east). Make a general statement about how the direction of sunrise and sunset differs between the December and June solstices.*

	Azimuth at Sunrise	***Azimuth at Sunset***
Mar. 21	__________	__________
June 21	__________	__________
Dec. 22	__________	__________

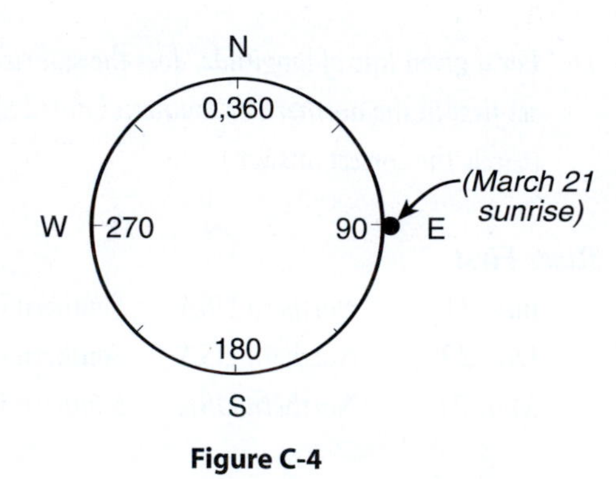

Figure C-4

Review Questions

So, why might a region be referred to as the "land of the midnight sun"?

How is the equator different from other latitudes with respect to seasonal changes in day length?

Appendix D

Weather Symbols

Following is a short list of weather symbols you may find helpful in completing some of the labs. The National Weather Service web site (http://weather.noaa.gov/graphics/chartref.gif) has a more complete list of the weather symbols. (The appendix of your textbook may also have a fuller listing—check there as well.)

Current Weather

- ● Intermittent rain
- ●● Continuous rain
- ** Continuous snow
- ,, Continuous drizzle
- ≡ Fog
- Thunder heard
- Thunder with intermittent rain

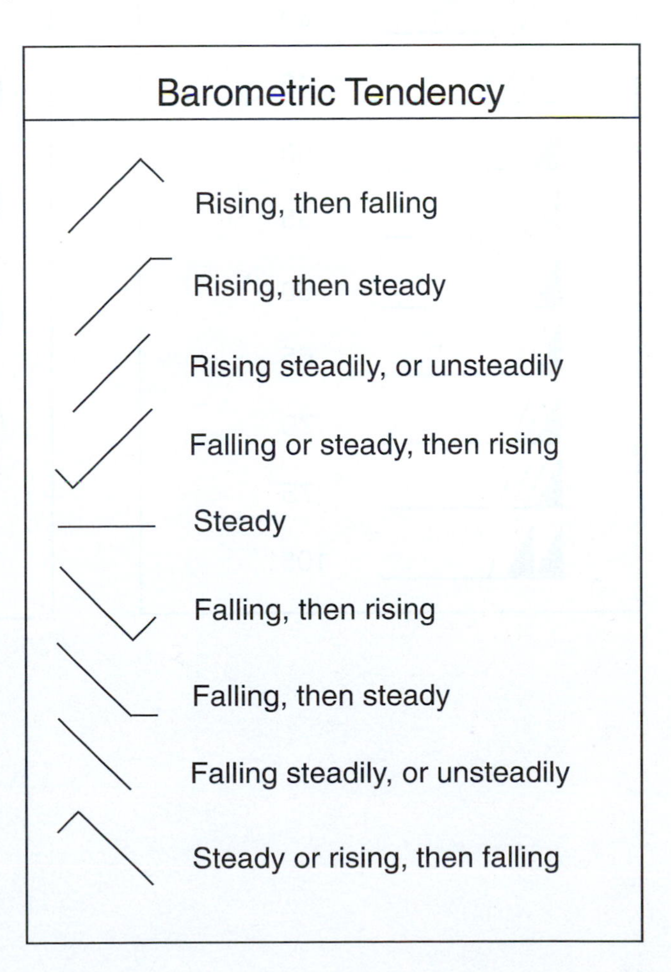

Winds knots

Calm
1–2
5
10
15
20
25
30
35
40
45
50
55
60
65
70
75
105

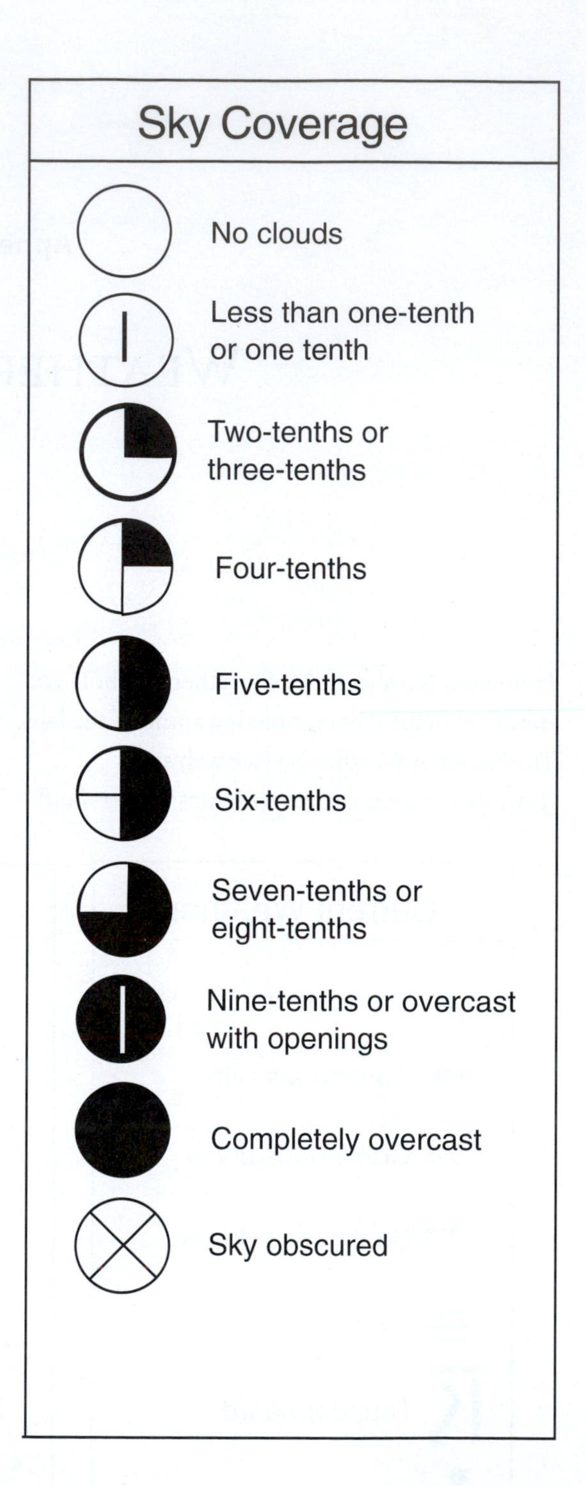